Building Shop

Building Shop

by J. Randolph Bulgin

Bulgin Forge Press
Franklin, N.C.

Building Shop
Copyright © 2008 by Randolph Bulgin
All Rights Reserved
Bulgin Forge Press
J. Randolph Bulgin
81 Bulgin Drive
Franklin NC 28734

ISBN 978-0-9785475-1-6

Printed in the United States of America
by Quebecor World Book Services

Introduction by George Bulliss
Design by Barbara McRae

George B. Bulgin — My uncle and the best machinist I ever knew.

The Craftsman

An ability to perform a task
May be learned by text
And rote repetition;
A craftsman's skill is learned
By passion for a task,
A quest for knowledge
And acceptance of the fact
That mistakes and faults
Are going to occur.

The ability to learn
From those mistakes;
The humility to share
Those mistakes,
The courage to risk
Making more mistakes,
Not intentionally, but
Through a learning drive,
Makes the difference in
Identifying an artisan.

By my son, George A. Bulgin

Acknowledgments

Thanking people who helped and/or supported in a project like publishing a book is always an "iffy" thing for me. I feel the support of the different groups of friends among the trade who gather from time to time at my shop or someone else's shop. I receive visits from time to time from old friends, I mean really OLD friends from back in school days and they are supportive. It would be impossible to name them all and I won't try to. But there are some who have to be mentioned.

Barbara McRae. She helped with my first book and she has been here for this one. Without her to work her magic on the stuff I come up with that is all it would be. Just "stuff."

My brother Fred Bulgin and my friend Fred Alexander. How would you go about a job like this one without a couple of Freds to call on when you need an extra pair of hands or a different lens?

Karen Wallace. A really smart librarian who can tell you where to learn almost anything about almost everything.

The Sunset Restaurant round-table discussion group who contributed absolutely nothing to this project.

And, as always and most importantly, Nan and George. Thanks. 'Nuff said.

Introduction

Down in the mountains of North Carolina, there is a small shop that has become a second home to many readers across the world. Through his articles in The Home Shop Machinist and Machinist's Workshop, as well as his previous book, Randolph's Shop, the author has given thousands a chance to share in his pride and joy, his shop. Once again, Randolph throws open his doors and invites the reader into the comfortable surroundings of his little kingdom. But be warned, you may catch the bug and find yourself in that endless (though arguably enjoyable) pursuit of building shop.

Randolph grew up around shops, spending his early days watching his father as he worked in his welding shop near the house. In the years that have passed, Randolph has not strayed too far from his roots, building his business and shop near all those early memories. Many men in Randolph's shoes, even those with far fewer years in the metal trades, would look back on their life's work and give themselves a big pat on the back for having mastered all the skills of their trades. Of course, once you reach the point in your life when you have convinced yourself that you know all the answers and there is nothing left to learn, you have sealed your fate. From that day forward, you will learn no more. There will be no more improvements in your work, no more advancements in technology for your shop, and the joys of shaving a few minutes off of a job, or finding a way to do what you thought could not be done, will no longer be yours.

It is fortunate for us that Randolph has not reached that point, and even better that he is not the type of man to ever reach that point. The joys of working metal, of conquering that impossible job, and building that elusive "perfect shop" will never leave him. It is fortunate for us because his love of the metalworking field comes through in the pages of this book, as well as his previous book and his many magazine articles. You may set this book down with plans for a new piece of equipment, or perhaps your wheels may be spinning with ideas on how to pry another ten square feet of shop space from that pile of household clutter that is your garage. You may even come away with some ideas to help squeeze a few more pennies out of every job. Regardless of what you take out of this book, you will be walking away with a smile; a little of Randolph's joy of metalworking is bound to rub off on you.

Whether an absolute beginner or a seasoned pro, those of us that choose metal as our medium to work our magic share some common traits. One of these traits is that endless pursuit of the perfect shop. A shop that lets us do all the things we need to, one that allows us to tackle the type of jobs we chose to do, in the most efficient manner. Randolph has come closer to building that perfect shop than many of us ever will. In this book, one of the best mentors a machinist could hope to have has opened the doors to his shop, his thoughts, and his projects. If you are not already there, after reading this book you may find yourself heading down that endless path of building shop. Don't worry, it's an enjoyable trip.

George Bulliss, Editor
Village Press, Inc.

Contents

The Craftsman, by George Bulgin   *vi*

Acknowledgments   *vii*

Introduction by George Bulliss   *viii*

Author's Preface   *x*

Chapter 1	Building Shop — A Philosophy and Practice	1
Chapter 2	Threading	13
Chapter 3	Rebuilding a (Your Machine Here)	33
Chapter 4	Debunking the Myths of the Gap Bed Lathe	45
Chapter 5	The Turret/Ram Milling Machine	55
Chapter 6	It's Never Too Broke to Fix	77
Chapter 7	Ornamental Iron — Again!	89
Chapter 8	If You Don't Want to Take the Job — Just Say So	99
Chapter 9	Building a Gun Safe	107
Chapter 10	The Care and Feeding of the Twist Drill	123
Chapter 11	Some Things You Need in Your Shop — and You May Not Even Know It	135
Chapter 12	The Engraver's Vise	147
Chapter 13	A Welding Quandary	163
Chapter 14	Turning Between Centers	171
Chapter 15	Machining with Soft Jaws	179
Chapter 16	Sweeping the Head of the Turret Mill	185
Chapter 17	Not Just a Lathe — A Contouring Lathe	191
Chapter 18	A Horizontal Milling Machine — On End	197
Chapter 19	The Wonderful World of Watts	205
Chapter 20	From the Chip Pan	209

Author's Preface

Writing a preface to this work is turning out to be more difficult than writing the book. And for good reason. I know how to weld but I am still learning how to write.

I have been engaged as a worker of metals in many — not all, but many — of its manifestations since I was a child. If I have paid any attention at all to the work I have done all of my life, I have gained some experience and some knowledge about the things this book is about. As it is popular to say now, "I have been there, done that, and have the T-shirt to show for it." But we are never as smart as we think we are — at least I am not.

After spending a lifetime, at least up to this point, doing what we do, it is difficult to admit that there might be something, sometimes a pretty basic something, that we don't know a lot about and maybe have never even heard of. We somehow take on the attitude that we should have knowledge about a subject and that if we don't it indicates some lack on our part and we are loath to admit to it. There is an expression I am particularly fond of quoting. I have no idea of its source but it goes like this:

Those who know not, and know not that they know not are fools. Avoid them.

Those who know not, and know that they know not are simple. Teach them.

Those who know, and know not that they know are sleeping. Awaken them.

Those who know, and know that they know are wise. Follow them.

I would probably add one more category to that and all of us can point to examples.

Those who know not but think that they know are dangerous. Have them arrested!

The point of quoting these philosophical homilies is this. I have had the opportunity over the years to work with people from all of the above categories. And the advice given about relating to each of them is good. It should be the responsibility of each of us to learn to recognize the types of people we come in contact with in our careers and to do as the saying suggests:

- Learn from those who know more than we do. And everybody knows something that we don't know.

- Teach those who know less than we do and who really want to learn.

- Ignore those who think they already know everything. But if you can stand to be around them long enough to find it, you might learn something from them.

- And try really hard to remember that you were once as ignorant as the youngest apprentice out there.

This book is an attempt to pass on some of the things I have learned in the past 50-plus years of working with machines and metals. Some of the lessons I have learned have left scars and those are the ones best remembered. Some lessons I have learned have cost me money. Those are pretty easy to remember, too.

This book is not so much about perspiration, as it is about inspiration. And hopefully it will also be a tribute to those who have inspired me over the years. My grandfather was a blacksmith who died in 1936, three years before I was born. My father was a blacksmith and a weldor who lived until 1995. And my uncle, who in his working years was the finest machinist I have ever known, is still living at the age of 91. All of these have inspired me along with many fine craftsmen I have worked with over the years. I will be happy if I know that I have done my part in some way to pass this information along to others.

Unfortunately it seems that there are now fewer "old guys" around to go to and ask questions of, now that I am one of them. And there seems to be a dwindling interest in this kind of work. I hope it is just that I now have a different perspective and that it just seems that way. But it worries me. It really does.

Randolph Bulgin

Chapter 1
Building Shop - A Philosophy and Practice

A six-inch cup brush mounted on a one horsepower hand-held grinder is not a tool, it is a lethal weapon! Give it a little leeway and it will take your clothes off of you!

I have a friend who visits my shop fairly frequently and on one such visit he said to me, "All you are doing is building shop." At first I resented the remark. I do much more than build shop here and can show you evidence of it in many forms but his comment made me think. And I realized that I do spend a lot of time "building shop." And as the idea grew with me it came to be not such a bad thing, building shop. It takes many forms and the type of work done in the shop becomes a side issue. After all, my shop is my hobby as well as my way of making a living. And if I want to spend my time making things to make other things then that is not altogether a bad thing. At least, not to me.

Several years ago, during the middle 1970s, I was in the business of making fireplace screens and accessories. We made shovels and pokers and the entire range of hand forged hardware that goes with fireplaces and there were five people, including my mostly retired father and myself, working in the shop. We all were engaged in the duties we were best at. My father spent most of his time at the forge. Two of the other men worked at their own welding tables, assembling the fire screens. There was one who I suppose would have been classified as an apprentice and his job was to keep a bunch of one-gallon sized coffee cans filled with the miscellaneous sawed bits and pieces required for the assembly of the screens. He also did most of the packaging and shipping. There were no problems about job classifications or about who got the good jobs and who got the bad. We all worked together pretty well and we got a lot of work done. During that time I designed and built eight different machines to do the specialized work of providing custom-fit fireplace screens to a wide variety of customers. What I did during those days which contributed most to the efficient functioning of that shop was — build shop. And I still do that today. I find time to make the parts and perform the repairs called for by my customers and of course I find the time to do the things that my wife plans for me to do. (I am not completely crazy!) The grass gets mowed here when it needs it, or at least it gets mowed within the next few days, and I get a lot of work done. But I cannot disagree that what I do the most of, and what I seem to enjoy the most, is add to the efficiency of my little machine shop by making new tools, by building new lifting devices, by building new work centers for my many specialized interests, and by continuously changing the way things are arranged. And then if it doesn't suit me I get the fun of changing it either to some other configuration or back to the way it was originally.

And that is what you are going to see in this book to a great extent. There is a chapter included here on building an engraver's vise and

I have also built an engraver's bench but I am really not yet a very good engraver. I hope one day to be but right now I engrave like I play the piano. I am not very good at it but I have a lot of fun trying. I built the vise and the bench because I could and because I had materials on hand which suited the purpose and because I really do want to learn more about engraving. But first I had to build a little shop! I always have to first do a little shop building.

A really important thing in shop building is having the material to build it from. When my brother and I were boys, 8 to 12 years old, I guess, we were constantly building something in our father's shop. And the search for suitable fasteners was more than a search — it was a quest. My father was working for the local power company at that time, this was during and just after World War II, and he no longer tried to serve the public from his shop. So the necessary stock of hardware and fasteners had dwindled. Our construction efforts were hindered because if it required bolting together there was a problem. It seemed that there were never enough fasteners of the correct size. Finding six or eight ¼"-20 bolts of the proper length, ¼"-20 nuts and ¼" washers and lock washers just didn't happen. If the bolts were the correct size the nuts had the wrong thread or the washers were too small. A typical job requiring that many bolts might wind up having two or maybe three sizes of fasteners and the length was never right. Because of those early years I cannot today pass a suitable washer lying in the gutter without picking it up and taking it back with me to place in one of my parts bins. 99.6% of the hardware I have stored in my back room will eventually end up in a land fill somewhere but that is alright with me because in the meantime I can rest easy that when I need a bolt or a nut or a snap ring of a certain size I likely have it. I have motors, gearboxes, bearings, electrical panels, casters, clevises, transformers, pumps, valves, air cylinders, hydraulic cylinders, filters, and probably three thousand pounds of just plain hardware in the back room of my shop. And that doesn't even include all the scrap steel, aluminum, brass, stainless steel, cast iron and synthetic materials I have stacked in corners. Keeping all of this "stuff" handy requires spending time arranging it so that I can find the right part when I need it and that involves what? You got it — building more shop! (Photos A and B)

Photo A. A small part of the shelves in the back room. By making the shelves shallow it is easier to see what is stored without having to move what is in front.

Photo B. More parts storage. The coffee cans are labeled otherwise I would have to look in every can every time I needed something.

But that is enough rambling about the philosophy of building shop and I long ago quit feeling like I have to explain to my friends why I do what I do. Let's look at some examples. Building shop isn't all pouring concrete and driving nails. Finding that piece of machinery you have spent years thinking about and looking for, and finding it at a time when you can convince yourself and those in your household that you can afford it, and learning that it is within range for you to hook up your little trailer and go get it, is a sublime example of building shop. Such it was with me when I found the Hardinge.

The Hardinge HLV-H

This little lathe is to the machinist what the Stradivarius is to the violinist. Other machines may also fit into this description. The Monarch 10EE has its following and deservedly so, and there are even today some imported machines whose features allow them to be considered as being contenders for this title. When I was a teenager and when Detroit was king of the automobile industry it wasn't uncommon to hear someone say, "I would rather be walking and carrying a Ford hubcap than to be caught riding in a Chevrolet!" Or maybe that was the other way around. But product loyalty was and is hard to figure out so I will stick with my original statement about

Photo C. The fabulous Hardinge HLV-H-DR. The Stradivarius of tool room lathes. This is the machine which I expect to be still here when all of my other toys have long since been disposed of.

the HLV-H. What else do you expect since that is what I now own? The lathe in the photograph you see here was in a shop in South Carolina. (Photo C) It had apparently been as appreciated and as cared for in its former home as it will be here in my shop because in spite of its being thirteen years old it was clean and showed no signs of the abuse so often seen in machines that age. The machine was purchased new by the machinist I bought it from and I even acquired all of the original invoices and manuals with it. You can always tell when a machine has been purchased by one guy and used by another. Nobody cares for a machine or anything else better than the one whose money paid for it.

So I went to South Carolina and brought the lathe home with me and situated it in the shop and immediately began building shop around it. I leveled it up and grouted it in. I built a chuck storage place for it. I built the work center with the tool holder lazy Susan on the top. (Photos D & E) I ran an air line to it and purchased a nice floor mat to go in front of it and now it is a part of my shop. And that makes it a part of me. As a shop building exercise a few years ago, I made several custom tool holders (Photo F) for an Armstrong wedge type tool post I had at that time. I sold the tool post when I sold the 14" lathe it was on but I kept the tool holders and happily they fit the Aloris tool post which came with the Hardinge. Now all I have to do is build a steady rest for it along with probably some sort of contouring attachment and a radius turning attachment for both convex and concave radii. Then there will have to be some sort of outboard support for turning long pieces through the headstock and of course I will need to make a knurling tool and — you see what I mean. I don't understand how people can *not* build shop! It keeps me busy even with my small shop.

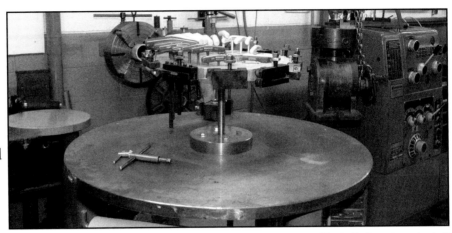

Photo D. A work center built especially for the Hardinge. It is conveniently located next to the lathe and the top is large enough to hold a reasonable number of tools while using the machine.

Photo E. Lazy Susan tool tray atop the work center. It accommodates 10 tool holders for the Aloris tool post and full sets of both Metric and inch Allen wrenches. I have had lathe jobs where 10 tool holders is not too many.

Photo F. Two of the tool holder blocks I made for an Armstrong tool post several years ago. They are machined from 4140 steel and heat treated to 42-44 Rc. You can't have too many good tool holders.

Building Shop

Photo G. The engraver's bench completed.

The Engraver's Bench

This is a wonderful example of building shop. In Chapter 12 there are plans for building an engraver's vise. There can be pretty firm plans or instructions for building a tool like the vise but the bench is a much more subjective and individual sort of appliance. A work center is, obviously, designed around the work you will be doing there. But equally important it must be designed around you and how you like for things to be. And last but not least, it must be designed around the materials and methods available to you when you build it. I had a circular piece of 1" thick steel plate lying outside my shop which came here as surplus in a trade I once made. I might have eventually needed it for some other job, maybe even a paying job, but it was too heavy to move around much in my storage area and it was going to eventually lose value if it lay out in the weather from now on so it became a candidate for use in my engraving bench. I also had as surplus the ¾" diameter

Photo H. Trying out for height by setting up on a barrel. Now is the time to make sure everything is cut to the correct lengths.

6 Building Shop

Photo I. Machining for the features on the bottom side of the table top. The overhead hoisting device comes in really handy here.

linear bearings for use in the mounting arm for the microscope and, with some modifications, the DC motor and controller which became the diamond tool grinding station. (Photos G, H, I, J, K)

Building this setup included before it was finished many side issues. Building a mount for the microscope which could be adjusted both vertically and horizontally had to be done. I wanted to be able to move the microscope out of the way quickly and easily yet be able to bring it back into location and focus without a lot of fiddling around with it. I needed the tool grinding station to be capable of accepting both ½" bore and 1-1/4" bore diamond grinding wheels without getting out wrenches and spending a lot of time in the changing. The tool holder itself required that it have the ability to produce,

Photo J. Wiring up the variable speed grinding spindle and the auxiliary 110V boxes. You will never have too many 110V outlets so I am adding 8 to the bottom side of the table. I might want to listen to the radio!

Photo K. The Lazy Susan with a magnetic chuck atop it. Really handy setup for engraving or for other operations which require holding oddly shaped parts for being worked on. Just because it is called an engraving bench doesn't mean that is what you have to do on it.

Photo L. The vise described in Chapter 12.

Photo M. Below, the tool sharpening fixture. Made from heavier than necessary materials for this purpose but I can also use it for sharpening wood chisels and larger tools.

Photo N. Right, another view of the tool sharpening fixture. The grinding arbor can accept other grinding wheels for use in sharpening many other tools.

accurately and with repeatability, the correct angles for the graving tool. The height of the vise in relation to the microscope and also in relation to the user had to be determined. And it is a good thing too, to catch gross errors in construction as you proceed so that you don't get nearly finished with the job and discover that something very basic is wrong and you have to nearly start over. I've been there, by the way! (Photos L, M, N)

The Lazy Susans may be used anywhere in your shop. They are easy to make, all you need is some scrap plate and a handful of steel balls. I find that getting up and down isn't as easy for me as it once was so if I can have things come to me instead of getting up and going to them it

Photo O. Sawing out the aluminum discs for the Lazy Susan bearing. Even hard wood may be used for these parts.

Photo P. Below, turning the groove for the balls. Use enough balls around the periphery to keep the top plate from rocking.

makes life a little easier. (Photos O, P, Q and R)

The accompanying photographs will tell you more about its construction than I need to talk about here. This bench and its built-in accessories suit me. I doubt if it would meet the needs of any of you reading this and, as a matter of fact, it may not suit me for very long. But when I outgrow it, or when it needs to be changed to serve other purposes then I will rebuild it. Or I will build another one but I will always be building shop. My wife doesn't understand why I am never satisfied with my

Photo Q. Lazy Susan completed.

Photo R. Showing the relative parts. The balls used here were ½" in diameter.

shop and I can't explain it. But I expect that the day I am satisfied with it will be close to the day I will leave it for the last time.

Dake Press

Here is yet another example of building shop. Some things come to you when you least expect them. All bad things come to you when you least expect them but we want to keep a positive spin on the things we talk about in this book so let's not go there. The big arbor press is an example of what I am talking about. (See photos S and T.)

An engineer of my acquaintance at a local factory called me one day and said that they were going to dispose of a big old press and if I was interested he would put a bid in for it. I went and looked and there was this Dake No. 4 arbor press. It looked pretty neglected. There were a few key parts missing but what was there was in really good shape and it looked like it deserved another chance at life. I won the bid at $51. I was able to make most of the missing parts and found that Dake still stocked spare parts for this press. Aside from the initial purchase, a bag of sandblasting sand and

Photo S & T. The Dake press as it came to me on the left and how it looks after a bit of TLC on the right. This is going to be useful in cutting those internal keyways.

a quart of paint, I spent a little under $400 on the machine and what you see here is the result. I now do almost all of my internal keyway broaching on this old fellow and it makes for pretty good company in my shop. I saw one sell on eBay for $1,600 just a few weeks after I finished with this one and it looked pretty rough.

7" Atlas Shaper

The last example I am going to talk about here isn't really an example of building shop but of adding to shop — and it all amounts to the same thing. (Photo U)

The shaper as a front line machine tool in today's shops is mostly a thing of the past. There are still places where it can be useful in a job shop from time to time but most of the time if there is still one in the shop it is covered up with dust and with anything which can be piled on it. I will never forget seeing a man running a shaper once in a shop where I worked. The material being machined was very hard. Somewhere in the 50s of the Rockwell C scale. The operator was a Greek immigrant (which has nothing at all to do with the story, by the way) and spoke very little English. But he knew his job and he knew what was required. The machine was

Photo U. The 7" Atlas shaper. This is more of a toy than a tool but remember how we look at this. All of our tools are toys and all of our toys are tools.

a 36" Cincinnati shaper with a massive vise on the table and the machinist was taking about a .800" depth of cut with a feed of about .025". He parked the ram at the back side of the part, heated the side being cut to about 1000 degrees (red hot) with a rose bud O/A torch and then took from 3 to 5 cuts. Then he parked the ram again, brought the work piece back up to heat again, and repeated the process. The shavings looked like those curly pieces of raw carrots you get in a fancy restaurant. The shop had a wooden tiled floor and there was a helper with a bucket of water and a cup whose job it was to run down the chips and cool them off after each cut so they wouldn't smoke up the place by burning into the floor. Some of you reading this will swear I am lying but the story is true and I saw it with my own eyes. I didn't understand much of what the machinist was saying in his native tongue but there was no mistaking the smile he had on his face as he did his job.

I doubt if I will ever do a job of that sort on this little shaper. I probably won't use it very much at all but the next time I have to make dovetail tool holder blocks for my Hardinge you may very well see it rocking away. I don't have to use it for internal keyways anymore because I have a Dake No. 4 arbor press.

Building shop takes on many faces. If shop is in your blood then it makes no difference how small or how inconvenient your place is. You will build a shop. I lived temporarily one time in an apartment. I had taken a new job and circumstances required that we not sell the house in the place we were moving from for a little while. I built a shop in that apartment. It took up a space of 30" X 30" and it even had a place to sit down. I couldn't take on much in the way of big jobs but I was able to reload my .45 ACP target ammunition and I also had room to set up my engraving vise — just not at the same time. If you are bound to have a shop then you will have a shop. Now if you will excuse me I have to get started building a steady rest for my Hardinge lathe.

Most of the time if a customer gets mad at me for some reason and says that he is never going to come back to my shop — it's OK with me.

Chapter 2

Threading

Threading

I considered as a title for this chapter the following:

Explaining the Processes Involved

in Machining Various Types of Threads

in an Engine Lathe Including Some of

the More Obscure Thread Forms and

the Methods Used in Machining and

Measuring Those Threads

But I decided to just go with the title: THREADING. It seemed to be a little less cumbersome. And I am told that it is not good writing practice to have a title which is longer than the text of the subject itself. This chapter first appeared in the October/November 2007 issue of *Machinist's Workshop*. After reading it back to myself I have decided that my explanation of threading on an engine lathe is no better than anyone else's but here it is anyway.

The chief drawback in asking someone what they think about a subject is that they will often tell you.

The process of producing threads is one that will be found in almost all machine shops. These days it has been transformed in many cases to entering a G92 in your machine control and then defining the parameters of the required thread. Lead, pitch, FPM (Feet Per Minute), DOC (Depth Of Cut), location of the thread on the part, etc. But threads and threading will always be here. In the job shop most of your threading requirements will be the simple production of a part which will have to fit another part. An external thread which will have to fit an existing internal thread or vice versa. The repair, or recutting of a thread will also show up on your menu from time to time. We will begin the discussion of this topic at a convenient starting place — the beginning.

SECTION A: Basic Threading

I spend a lot of time lurking on one or another of the many metal working internet forums and aside from discussions about the bargains or the lack of bargains to be found on eBay, and the inevitable comparisons of one imported machine tool to another, there seems to be more discussion and questions from the novices about threads and threading than almost any other facet of the machinists trade. And with good reason. I machined my first thread on an engine lathe using a single point tool, which I ground myself from a HSS (High Speed Steel) blank, nearly fifty years ago and I still don't know everything there is to know about it. But I know a lot more than I did back in Machinery Repairman "A" school at the United States Naval Training Center in San Diego, CA in 1959.

I fully understand that doing something for a lot of years does not make one an authority. So I suggest that you take what I am going to say here as it is intended. I am going to relate to you some things that have worked, and some that have not worked, for me over the years. I will be just the next in a long line of self-proclaimed experts who is willing to share some of my experiences in the machine shop and you can take this for what it is worth. Like most free advice, the value is commensurate with the cost. And I will add one further caveat. The approach I am taking here will bore some of you to tears. I am really going to try to talk to the novice who has never fully understood the thread cutting process. The rest of you will either just have to bear with me — or ignore me altogether.

I was taught in high school physics class that there are three basic simple machines. Those are the lever, the wheel and axle, and the inclined plane. All of the more complicated mechanical movements are combinations of, or variations of, these three basic mechanisms. The thread is an application of the inclined plane. For those of you who have an interest in historical trivia there are many sources of information available about the development of the various thread systems recognized today. It can be a particularly interesting subject in that it includes the history of the development of our modern thread cutting engine lathes. But that is not the subject of the first portion of this chapter. We are going to talk here about the basic, the very basic, process of cutting threads on a lathe with a single point tool.

Let us talk a little about the lathe first. If you are interested enough in the subject to have read this far you will already know what an engine lathe is and perhaps you even own one. But owning a machine and having a complete understanding of it are sometimes two different things. (I own a computer if you want an example!) The thread cutting engine lathe does two things. It rotates the work and it moves the cutting tool along the work at a constant rate so as to produce the desired thread lead. If your lathe has a lead screw with 4 threads per inch and the thread you want to cut has 16 threads per inch, you will have to change the gears between the spindle and the lead screw so that the spindle makes 4 revolutions while the lead screw makes 1 revolution. If you need to cut 24 threads per inch — 6 turns of the spindle to 1 turn of the lead screw. You see where we are going here? The pile of gears underneath the workbench, which seemed always to be missing the gear you were looking for, is fading into history and I, for one, will not miss them. But knowing

Threading

Photo 1. Thread cutting tools:
Top row left to right:
Iscar SER 0625 H16
Kennemetal NSR-123B
Kennemetal KER-10
Brazed carbide ER-10
Brazed carbide 87 755

Bottom Row left to right:
HSS for #3 Acme
HSS for fine thread close to shoulder
Thread gage for 60 degree threads
HSS for #5 Acme
HSS for small bore internal thread
HSS for #8 Acme

the relationship between the rotational speed of the spindle and the rotational speed of the lead screw is a good thing. Let's just not get bogged down by it. Most of the lathes we see today, even including the smaller machines found in basements and in garages, have quick change gear boxes so that it is relatively easy to make these selections but it does help to understand what goes on when you change the position of the levers on the outside of the gearbox.

Now let us consider the cutting tool. **Photo 1** shows a variety of thread cutting tools. This is just a small sample of the configurations of tools for cutting threads we have available to us these days. I don't recommend one as being in any way superior to another. It just has to do the job and all of these tools have done the job for me at one time or another. The hand ground HSS cutting tool is going the way of the change gears for the lathe. Sintered carbide inserts are becoming so effective and readily available that spending the time grinding cutter bits, and burning your fingers in the process, is becoming a thing of the past. But — like understanding what the gearing does, knowing how to grind a tool bit and knowing the features necessary in a properly ground cutting tool is good. The variations of rake angles, clearances, proper shape and all of the facets of a properly prepared cutting tool are something that you need to know about. And it is a great subject for another discussion but we are going to keep focused here on the actual process of cutting the thread.

The example I am using here is as simple as I can make it. We will cut 16 threads per inch on a .750" diameter part. There are no distractions. We will not be cutting up to a shoulder. We are not cutting threads per inch on a Metric lathe or vice versa. No internal threads. No pitch diameter measuring tools will be needed. We are simply going to cut a thread and make it fit the nut which we will be using as a gage.

Start out by preparing the blank. I am machining here between centers but that isn't a requirement. If you do cut threads between centers, be sure that the backlash between the lathe dog and the driving plate or chuck jaw is always removed. Also, if you remove the work from the machine during the process for any reason, make sure you put it back into the machine with the same relationship to the driving plate. For purposes of this exercise I have relieved the end of the thread nearest the headstock to provide an easy stopping place.

Self confidence — the practice of ordering exactly enough of an exotic material to make all 75 parts with nothing left over for spares.

Set up the machine for threading. And here we will deal with the first major misunderstanding many novices and some experienced machinists suffer from. The cutting speed or FPM for thread cutting. Cutting speed is cutting speed. The recommended cutting speed for single point machining of 1018 steel using a HSS cutting tool is about 90 FPM. With a carbide insert it is 300 FPM or even greater. This does not change just because you are cutting threads instead of turning a diameter. But there is another factor to consider here. If you use a carbide threading insert and try to run at the recommended speed, your part will be turning at a little over 1500 RPM (Revolutions Per Minute) and the tool will travel the entire length of the job in about 1-1/4 seconds. Getting a half-nut engaged and then disengaged at the proper locations will be a little exciting, to say the least. So we must compromise. The configuration of this part allows for a higher speed than many threaded parts because of its easy access at both the starting and the stopping ends. But we still have to consider reaction times. I machined this thread at 490 RPM.

The next part of the setup is to set the

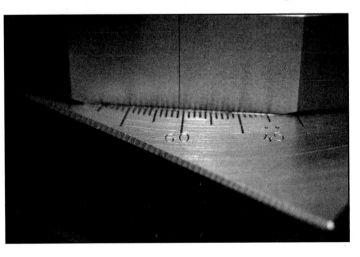

Photo 2. Compound setting for 60 degree threads.

Threading

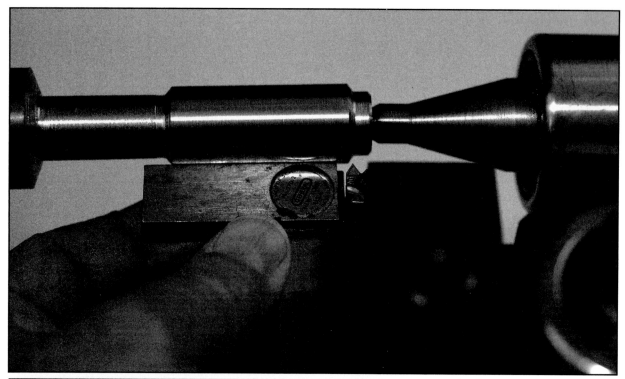

Photo 3. Top: Checking tool to be square with work
Photo 4. Below: Setting tool on centerline.

cutting tool relative to the work. Begin this process by setting the compound rest to an angle of 60-1/2 to 61 degrees from the spindle axis as seen in **Photo 2**. The desired angle is 29 to 29-1/2 degrees from the perpendicular. Here is another point frequently taken for granted and not explained to the novice machinist. We are cutting along the flank of a 30 degree angle so why not set the compound at 30 degrees? A good and valid question and here is a good and valid answer. Since we are feeding parallel to that angle in small increments there will be a series of marks on the flank of the thread where each cut is taken. By setting the compound to a lesser angle we can take the cut with the leading edge of the tool while the back side of the tool will have a scraping or cleaning effect on the back side, or trailing side, of the thread.

Now shift the gear selection levers to the positions indicated for machining 16 threads per inch. Just as an example, the chart on the headstock of the lathe I am using, a Kingston HJ-1700, for this job calls for settings of LBS1. That means that there are four levers or dials which must be positioned in order to cut this thread. One lever is set, as indicated, to L, one to B, one to S, and the rotating selector dial is put at 1. If you are using a different lathe from the one I use then obviously the settings will be different but the result will be the same. Also on most modern machines there is a safety feature which prevents engaging the half-nut while the feed clutch is engaged and vice versa. This control will be on the lathe carriage in the form of a lever or knob. Set it to the required position for threading.

Photo 5. Compound set to zero.

The next setting to be made is to position the mounted cutting tool and compound rest relative to the work piece. The cutting tool is positioned in the tool holder or tool post so that it is perfectly square with the axis of the work piece and is aligned vertically on the centerline as shown in **Photos 3 and 4**. (In **Photo 4** a machinist's scale is trapped between the work and the tool point. When the scale is vertical, the tool is centered.) Next, set the dial on the compound rest to zero. See **Photo 5**. Then move the point of the tool, using the cross slide, until it just contacts the work. If the work is rotating, move the tool point until it makes a barely visible mark around the circumference of the part. At this point you should set the crossfeed dial to zero and if you have a DRO, (Digital Read Out), set it to zero also. This is the point to which you will return the cross slide for each cutting pass.

Photo 6. Threading dial on Kingston lathe.

Threading

Photo 7. Checking threads for correct lead using a thread gage.

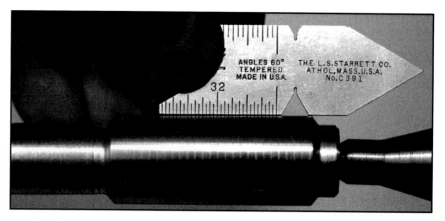

And last but not least, move the carriage by hand all the way along the required travel of the threads to insure there will not be an unexpected and unpleasant interference between the compound rest and the chuck jaws. If you chew tobacco spit out your cud at this point so that you will not swallow it at an inconvenient time. Now cut some threads.

Photo 8. Cutting pass in progress.

Move the point of the cutting tool to the right, or tailstock end, of the workpiece until it is about a half-inch or so past the end of the part. Move the cross slide to the pre-established zero point and then advance the compound rest by approximately .002". Here is where you will make your first cut but here also is a good place to review for a moment the workings of the thread cutting dial. **Photo 6** is a picture of the thread cutting dial on my Kingston lathe. There are many other variations. Here is what I consider to be a fair summation of the use of the thread dial:

1. For threads whose lead in TPI, (Threads Per Inch), is a multiple of the TPI of your lead screw, i.e. 4, 8, 12 16, etc., engage the half-nut at any point on the dial.

2. For threads with an even number of TPI, i.e. 2, 6, 10, 14, etc., engage the half-nut at any of 8 equal points around the dial.

3. For threads with odd numbers of TPI, i.e. 3, 5, 7, 9, etc., engage the half-nut at any of 4 equal points around the dial.

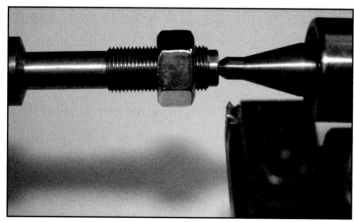

Photo 9. Finished thread.

20   Building Shop

4. For half threads, i. e. 2-1/2, 3-1/2, 6-1/2, 11-1/2, etc., engage the half-nut only at 180 degree points around the dial.

I am probably going to hear from people about the preceding but it works for me and has for a long time. If you are unsure of where you should engage the half-nut you cannot go wrong by engaging at the same point around the dial every time. Just remember that the thread dial is your key to not suddenly and unexpectedly splitting the thread on a nearly completed job. This usually results in a lot of unnecessary profanity and tool flinging and should be avoided whenever possible.

Now back to the job at hand — cutting these threads. Since the threads on this part, 16 threads per inch, is a multiple of the 4 threads per inch on my lead screw, I can engage the half-nut at any point and be sure I will enter the thread at the same place every time. A good habit to develop, however, is engaging the half-nut at a point you have chosen rather than just throwing the lever. Stay in control of the machine. Engage the half-nut and take a first cut, as we have already mentioned, at a depth of .002". At the end of the cut disengage the half-nut, back the cross slide away from the work, return to the longitudinal starting point and move the cross slide back to zero. Now check your setup with a scale or a screw pitch gage making sure that all of your machine settings are correct. See **Photo 7**. If everything checks out, proceed with machining the thread. I will usually feed in up to .005" per pass for the first 3 or 4 passes and then reduce the amount of infeed gradually until I am taking from .001" to .002" per pass. It depends to a great deal upon the material and the cutting tool. **Photo 8** shows a cutting pass in progress. Notice that the chip is being removed by the leading edge of the tool. When you are approaching a point where the thread begins to look finished, pause between each cut to check with whatever you are using for a gage, in this case a standard ¾" — 16 TPI nut. As you near the completion of the thread, advance the compound in .0005" increments until the nut will screw on to the thread without using a wrench. It should run freely but without any "wiggle" or "slop". **Photo 9** shows the nut on the finished thread. De-burr the start and stop ends and you have a threaded part.

I began keeping a notebook many years ago about operations performed in my shop and it included information about each thread I was asked to machine. I still keep and add to those notes except that I rarely machine a thread these days that I have not at one time or another machined in the past. The most useful note is the amount of infeed I make on the compound for a given thread. In our example for this chapter my notes tell me that I will have to feed in a total

I try to be impartial in the service I give my customers but the person who cleans the grease off the job before he brings it and pays me promptly after I finish it will usually get a little preferential treatment.

of .0505" on the compound in order to complete the thread. I start checking with the gage at about .048". It can vary by a couple of thousandths depending upon the material, the outside diameter of the part, the condition of the gage, the condition of the cutting tool, and maybe, even upon what you had for breakfast. But it should run about the same each time you cut the same configuration of threads.

This is just a start as far as the single point machining of threads on an engine lathe goes. But it is a good start. And if you master the process we have discussed in this section then other, more complicated threads and threading will come easily. If you stay in the business of machining, whether for a hobby or for a living you will be called upon to do a lot of threading. Some of the complications or variations will include:

- Machining Metric threads on a lathe which is equipped with an inch lead screw and vice versa.
- Setting up on an existing thread to re-cut or to make repairs.
- Threading a part for which you have no gage. Measuring pitch diameter.
- Multiple lead threads.
- Cutting close to a shoulder.
- Internal threads in a blind hole.
- Other profiles. Acme, square, buttress, Whitworth, etc.
- Left hand threads.
- Tapered threads.
- Combinations of all of the above.

All of these and more will come your way eventually and all are good subjects for a more in depth discussion of threading on the engine lathe. We are going to visit some of them later on. In the meantime, my apologies to those of you reading this who already know all there is to know about the single point thread cutting process. But maybe some novice, somewhere, will gain something from the discussion. And then they can pass the information along to someone else. Isn't that the way it is supposed to work?

SECTION B: Metric vs. Imperial

It will be long after I am dead and forgotten when the global unification of measuring systems finally happens. If it ever does. I

A friend is one who can call attention to your shortcomings and you won't get mad at him. Or at least, you won't stay mad long.

am not even going to talk here about whether one system is superior to another although I have some opinions about the subject. One thing is certain, however. Trying to design or manufacture machined parts using two different systems at the same time is a sure recipe for disaster. Remember when NASA missed a planet or some other heavenly body because somebody used miles instead of kilometers? I worked for several years in an environment where the maintenance machine shop was responsible for the upkeep and repair of machinery from all over the world. There was a rolling mill built in Germany and equipped with Japanese motors and controls and American built hydraulics which had every system known to civilized man at the time of its manufacture. And a few which seemed to me to have originated in uncivilized parts of the world. Our tool crib was an organizational challenge which never was perfect in spite of some good people trying to manage it. I will include one example here:

- 1" NPT pipe thread, according to Machinery's Handbook, has 11-1/2 threads per inch and an outside diameter of 1.315".

- 1" British Standard pipe thread, quoting the same source, has 11 threads per inch and an outside diameter of 1.309".

Even a practiced eye has trouble distinguishing between these two thread forms yet when they are used in extremely high pressure hydraulic systems, greater than 3000 PSI in many cases, using the incorrect thread form can have disastrous consequences. Is it any wonder that people not familiar with our line of work think we are crazy?! It isn't even 1" for Heaven's sake! And this is just one tiny example from tens of thousands of individual parts the machinist must deal with in a maintenance shop or a job shop environment. But the fact remains that the systems are here and must be dealt with. (Boy, if I could just be king for a couple of days!!)

Metric threads are machined on the lathe just like Imperial threads are. But there is one major difference. If you have a lead screw on your lathe which is measured in TPI and you have to cut a thread with a lead measured in millimeters, or vice versa, there will be a problem. Not an unsurmountable problem but a problem, never-the-less. Many lathes made and sold today have gearing in the quick change boxes which will produce both series of threads but the gears in each case will be driving a single lead screw and it will be one or the other – not both. Your thread cutting dial will be useless. Here is the solution and it works both ways, inch threads on a Metric lathe or Metric threads on an inch lathe:

You must lock the lathe carriage into lead and not take it out until you are finished with the thread. Engage the half-nut and when you reach the end of the thread, stop the spindle, leaving the half-nut engaged, back your cutting tool out to clear the workpiece, and reverse the machine to bring the carriage back to the starting point. (If your lathe has a foot brake on it you will appreciate it here.) Repeat this process until the thread is complete. If you accidentally disengage the half-nut you will have to reposition the cutting tool using the process for repairing threads which we will discuss next.

SECTION C: Re-Machining Or Repairing Threads

This section describes a process you will be faced with many times if you stay in this business long enough — setting up the lathe to make repairs to, or re-machine, an existing thread. It is one of the operations which can conveniently be done in a job shop or maintenance shop no other way than in the lathe. You will be faced with the challenges of doing this many times and for many different reasons. One is the repositioning of the cutting tool when cutting Metric threads on an inch lathe — the subject we were just talking about. Other reasons for having to "pick up" an existing thread on the engine lathe can be:

- Resuming thread cutting after having to replace the cutting tool for some reason.

- Making repairs to a threaded part which has been damaged or mutilated.

- Providing a looser fit to an existing part.

- Or maybe you had to remove a partially completed thread from the lathe to take on the "emergency" job the foreman just handed you. (You need to have a few words with that foreman if this is the case!)

Photo 10. Left, a 1.376" OD X 18 TPI thread which has been damaged. Photo 11. Right, moving the cutting tool into position for repairing the thread. Note that the tool here is out of alignment with the thread.

Photo 12. Threading tool in position to re-cut the thread.

Photo 13. The completed repair. Notice here that the damage to the thread is still evident but the nut travels past it without interference.

Whatever the reason the time will come when you will have to do it and here is how. First, make sure of your machine settings as you always do. Select the proper gearing, set the compound to the correct angle for the thread you are working on, position the tool relative to the work piece, hold your tongue just right, etc. Then with the tool withdrawn to clear the part in the X axis, start the spindle and engage the half-nut. When the tool has moved past the start point of the thread, stop the spindle, leaving the half-nut engaged. Now, using a combination of the cross slide and the compound rest, move the point of the tool into position. Turning the spindle by hand or jogging the spindle if you have a jog feature on your machine will help in this process. **Photo 10** is a picture of a part threaded to accept an N07 locknut, 1.376" in diameter with 18 threads per inch. It runs against my grain to do it, but I have intentionally damaged the thread with a hammer as can be seen in the photograph. **Photo 11** shows the tool out of position relative to the thread and **Photo 12** was taken with the tool properly positioned for re-machining the thread. **Photo 13** is a picture of the nut back on the repaired thread. Notice that the hammer mark is still visible. This process will remove the misplaced material which interferes with the thread but can do nothing to put material back. I suggest that the best way to overcome this problem is not to hit the thread with a hammer in the first place but we all know about millwrights and hammers.

Threading

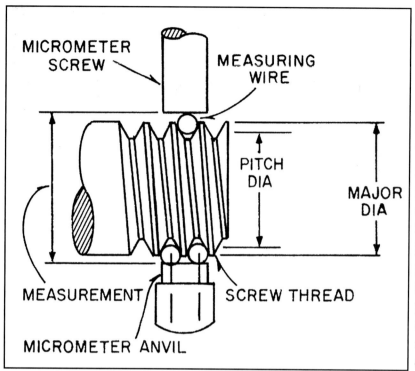

Photo 14. A drawing showing the method used for measuring the pitch diameter of a thread. Taken from a U.S. Navy training manual.

SECTION D: Threading A Part For Which You Do Not Have A Gage

Now we are coming to the place where we separate the men from the boys in the machine shop. This isn't really that difficult but it is one of the places where a good customer will appreciate a good machinist. You will inevitably be presented with the job of machining a threaded part and you will not have a gage to try before removing the part from the lathe. I could list a thousand examples but I will name only one. There is a dam on the Little Tennessee River near where I live and one of the valves which turns the water into the turbine is a gate valve with a 3" diameter, #3 left hand Acme thread valve stem. The valve was installed in 1928 and had been in place for nearly 50 years when the internal threads in the valve bonnet finally wore out. There was a replaceable bronze nut in the bonnet and the threads were completely

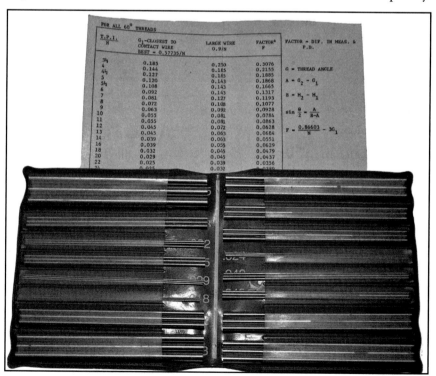

Photo 15. A set of thread measuring wires and the chart which goes with them.

26   Building Shop

stripped out of it. My job was to machine a new bronze insert. There was no possibility of removing the valve stem for use as a gage and, even if it could have been removed, it was much too long to bring to the shop. So I had to improvise and if you stay in this business for very long you will too. Here is one of the places where you will find that your *Machinery's Handbook* will be worth what you paid for it.

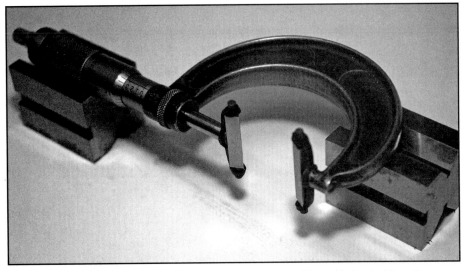

Photo 16. A set of thread measuring triangles mounted on a 1"–2" micrometer. These triangles are simpler to use where possible because of their relative ease of handling.

I will not go into an explanation here of the step by step use of these measuring methods. The principles will be obvious and the measuring devices themselves come with adequate instructions for their use. But you should become familiar with the various methods as some will lend themselves to some jobs where others will work better in other applications. Whichever method you use it is important to remember that your cutting tools must be precisely shaped to produce the correct thread form. All of the measuring systems mentioned here depend upon an exact thread configuration.

External threads are, of course, the easiest. And the best known method for measuring external threads is the three wire measuring system. **Photo 14** is a drawing I have scanned from a U. S. Navy training manual from a long time ago. But it clearly shows the method for measuring the pitch diameter of a thread using three wires. This system can be used for almost any thread form except square. In **Photo 15** you see a set of thread wires I keep in my tool box while **Photo 16** shows a set of thread triangles I will frequently use. The triangles are sometimes a little handier to manipulate due to the rubber bands made for attaching them to the micrometer. Handling a threaded part, three small wires and a micrometer all with just two hands will sometimes challenge your dexterity and you may be calling for an extra pair of hands. All of the thread measuring systems described here will have formulae with them and whichever method works best for you in the application at hand will be the best.

Another method, and the one I used in machining the #3 left hand Acme I described earlier, is to use a gear tooth vernier. It is shown

Threading

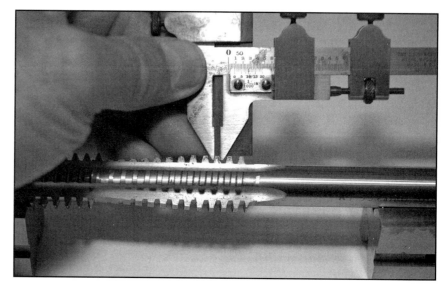

Photo 17. Measuring the thread profile on a #6 Acme tap using a gear tooth vernier.

Photo 18. Thread gages, both a ring gage for gaging external threads and plug gage for gaging internal threads.

in **Photo 17** being used, for the purpose of this demonstration, to measure the form of a #6 Acme tap. This method is most useful for the larger sizes of Acme, buttress and other similar threads.

There is one more point to make about reproducing threads and that concerns internal threads. When you have to machine an internal thread to a drawing or to fit a part unavailable for fitting, you must first make for yourself a gage. You must first machine the matching external thread, using one of the systems described here, then machine the female thread to fit it. **Photo 18** is of a gage I made to fit a 2.612" OD X 8 TPI part. I don't even remember what the part was but it was something else for the local power company.

SECTION E: Multiple Lead Threads

The first multiple lead thread, sometimes referred to as multiple start thread, I was ever called on to machine was a valve stem which was used to isolate the sight glass on a Navy ship's boiler. The thread was a 4 start thread with a pitch of 8 TPI, notice we call it pitch here and not lead, and was designed to be quickly closed in the event the sight glass was broken. A good example of multiple start threads may be found on the cap of a plastic milk jug. They provide fewer turns of the nut or of the valve wheel or even the cap on the toothpaste tube. This is where the difference between pitch and lead is important. In a single lead thread, pitch and lead are the same and are often used as interchangeable terms. But in a multiple lead thread they are different. In the example being used here for purposes of demonstration, we are cutting a thread with a pitch of .125" but which has a lead of .5". Let me show you what I am talking about.

To produce this example proceed as if you were going to cut a standard thread of 8 threads per inch. Machine your blank to the proper diameter, set the compound rest at 29 or 29-1/2 degrees, and zero the compound and the cross slide. Set or adjust all of the tool/workpiece relationships we have discussed earlier. But there are a couple of important differences and the first one is the configuration of the cutting tool. There is an additional clearance which must be considered. The helix angle on this thread is much steeper therefore the shape of the nose of the cutting tool must be changed in order to provide clearance between the tool and the flank of the thread. This is called, strangely enough, flank clearance. **Photo 19**, taken looking straight at the nose of the tool, shows the angle referred to. Most manufacturers of threading inserts now include tooling, inserts and anvils or shims, for almost every conceivable thread configuration and helix angle for either right hand or left hand threads. This is what helps to make tooling catalogs thick and heavy!

The second major difference in this setup is gearing. The machine must be geared to produce 2 threads per inch — which we will do 4 times. Machine the first thread using figures for an 8 TPI thread, i.e. feed the compound in .102" (from my notebook) and then return the compound to its zero setting. See **Photo 20A**. Now remove the workpiece from the lathe, rotate it exactly 90 degrees, and machine the thread as shown in **Photo 20B**. Repeat the process until you have the completed thread seen in Photo 20D. I accomplished the exact rotation, or indexing, of the part by machining the part

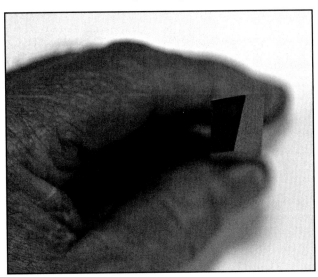

Photo 19. A pronounced flank relief angle ground on the nose of a HSS threading tool.

Threading

Photos 20A through 20D ---- The four steps in producing a four start, 8 pitch thread.

20A

20B

the exact rotation, or indexing, of the part by machining the part between centers with the headstock center mounted in the 4-jaw chuck and using each of the four jaws to drive the lathe dog in turn. Another method, and probably the best for cutting internal threads when you cannot work between centers, is to set the compound rest parallel with the lathe axis and advance the tool point using the compound feed crank by the amount required, in this case .125", between cuts. If you do this you will have to infeed using the cross slide and the tool will be cutting on both sides.

Threading

20C

20D

One last thing to note here is that the lathe carriage will be moving 4 times as fast along the axis of the work so you will probably have to adjust your speed accordingly. I machined this example at 110 RPM.

SECTION F: Other Difficult Threading Problems

The remaining items on the list of "different" or "out of the ordinary" threading issues on the lathe are mostly combinations of the ones we have been talking about. Cutting threads up against a close shoulder will probably involve some kind of a tragedy before you master it. The same goes for cutting internal threads

Threading

in a blind hole but there are solutions. Mounting the cutting tool upside down and machining the thread in reverse is a trick often used. Be careful if your lathe has a threaded spindle, by the way. Machining different thread profiles is simply a matter of acquiring and using a tool of the correct shape. Left-hand threads and tapered threads are not difficult — just different. Machine tapered threads using the same methods you use to produce tapered diameters — tailstock setover or a taper attachment. There are tricks to the process of thread cutting, some of them good and some of them worthless, and I won't even try to tell you that I know them all but I have learned some of them over the years. The best trick I know of is just to take the jobs as they come, use your intelligence and your experience as you approach them, and pay attention. And don't believe everything that somebody like me tells you!

CNC machines have so greatly simplified the process of producing special parts that the art of making them on the conventional lathe is fading almost as fast as us old machinists are, at least in this country. But the need will be around for a long time and if you master it you will have a marketable skill. And a sense of satisfaction at being able to do something not every body can do.

Craftsmanship doesn't know the difference between metal, wood or clay. If it is present it will show whatever the material.

Chapter 3
Rebuilding a (Your Machine Here)

Rebuilding a (Your Machine Here)

Working with used machinery is a complicated thing which depends to a great degree on just exactly what is expected from the machine after it has been purchased and moved to your shop. The machinery dealer wants to make it as appealing to a potential buyer as he can without investing more than he will likely be able to get out of it. The working shop owner wants the machine to be put to work. Unload it, set it up, hook it up, and put it to work. The collector looks for perfection and will want to make the machine as nearly perfect as he can. If we think of the used machinery market, or at least the purchasers of used machinery, in terms of being classified from One to Ten, I would put the guy who just needs a machine to go to work for him as being a One while the collector would be a Ten.

All that is required by a One is results. He will unload the machine in his shop and hook it up and turn it on. If there are broken or missing parts they will be repaired or replaced as he comes to them and if the machine will work without that part then don't worry about it. The Ten, on the other hand, will likely disassemble the machine completely, evaluate every part of it for wear and overall

Photo 1. The machine as it was when I unloaded it. Complete but in pretty much of a mess.

condition, replace any worn fasteners, repaint or refinish every component and wind up with a machine as good as, or in most cases better than, brand new. Most of us will lie somewhere in between a One and a Ten.

The machine being used here as an example is going to get what I would define as about a Seven or maybe an Eight treatment. I have had this Dake/Johnson horizontal band saw for a couple of years now and I am finally getting around to working on it. I already have a serviceable cutoff saw so I don't need to rush the machine into service. It is a higher capacity machine than the one I am using now so I will probably switch the saws after I get it put back together. For that reason I want it to be put into first class working condition but I am not going to enter it into any "machine rebuilding contest" so I can take some shortcuts in the process. Maybe it will even be as low on the scale as a Five or a Six but I plan for it to be a good working machine when I get finished with it.

Photo 2. Beginning disassembly. Wear old clothes for this part of the job.

This particular machine is one which I happened on at an auction. I don't pretend to be an expert on machinery auctions. There are some experts around and they usually won't share their expertise nor the benefit of their experience and that is OK. But there are some things I have learned from observation at some of the auctions I have been to and since I am not an expert I will share. The auction where I bought this machine was an auction of wood working machinery at a high volume manufacturer of office furniture. The factory was being closed and the machinery and inventory were being sold. The auction was well advertised and there were many potential buyers present. The good ladies of the local Methodist Church were there to sell hotdogs, cold drinks and hot coffee and it was a real event. But here is what turned out to be the good thing about it for me. Most of the potential buyers were also manufacturers of furniture or other wood products. They knew a lot about wood working machinery and about

The more things a single machine is proclaimed to do, the poorer the job it will do of any of them.

Photo 3. Getting down to the bare bones.

Photo 4. Don't disassemble all of the sub-assemblies at one time. Unless you are blessed with a better memory than me. And you probably are.

the large inventory of raw materials available. That was their specialty and they were knowledgeable about it. But they were not looking for Dake/Johnson horizontal cutoff saws or Darex drill sharpening machines nor Whitney Model 92 bench-mounted punches and I bought one of each that day for pocket change. It doesn't happen at every auction but I am told that once in a while even a blind pig will find an acorn in the leaves.

Regardless of where or how you may come into possession of a used machine, of whatever type, there are some things universal to the reconditioning process and that is the subject of this discussion.

The first thing to do is, whenever and however you acquire a machine, always ask about a maintenance manual or a parts list. You will be surprised how many times the casual question, "By the way, do you happen to have the manual for this machine?" will be answered with, "Oh, yeah. Now that you mention it, I think I do know where that is." It won't happen every time but you never know. Sometimes there may even be a maintenance history included which can be useful. It depends too, upon the machine. A maintenance manual for a 340-pound anvil, for example, will

Photo 5. And now out to the sand blasting pad. Going to be a long day!

be of marginal value. But as the machines get more complex the value of a good parts and operation manual goes up. The manuals for this saw were not available where I bought it but I did find them on the Internet and was able to print them in my office.

The first thing to do after you get the machine back to your shop and get ready to go to work on it is to take a lot of pictures. Take pictures of the machine from several angles paying particular attention to the parts or sub-assemblies which will be disassembled during the reconditioning process. And continue to take pictures as you work. Digital cameras have come into their own and this is a good place to put one of them to use. I have a friend who has done a lot of in-depth rebuilding of machinery and he is blessed with a memory which I do not have. He can disassemble a machine and, months later, reassemble it and remember which spacer goes on the shaft on which side of the bearing. I cannot do that. I have to have photographs or drawings and the pictures I take provide me with that. And if at some future time you decide to write a book about it you will have the required photos already at hand.

It is also a good idea to take notes as you go. When you start reassembling the machine and you have plenty of reference photographs and a manual and your notes to back them up you

Rebuilding a (Your Machine Here)

Photo 6. A pleasant job. Not!

will see that, "Oh, yeah. The frappenschlitzer flange goes under the dorkenporken collar instead of on top of it." And reassembly will go more smoothly.

How far you go in the process will depend to a great degree upon what resources you have available in terms of your own facilities. Do you have what it takes for sandblasting? Are you reconditioning the machine for resale or will you keep it and use it yourself? Do you care if the old paint is left on and painted over? Do you care if it is painted at all? All of these questions and more will have to be thought of and decided upon as you undertake the job.

My advice is this. Do the job as if your high school shop teacher is going to grade you on it. Balance the expense of having the saw table or the vise jaws reground on a Blanchard grinder against having to look at the scars on the table that you didn't even put there. Granted it does not make sense to spend a thousand dollars on rebuilding a machine for which you paid two hundred dollars at auction and which, when finished, will be worth a maximum of eight hundred bucks. But don't take short cuts which will make your work look less worthy than it is. You will have to be the judge of where the practical limits lie.

But now let's get busy with the job at hand. Begin the disassembly process by removing all of the identifying tags and labels you will want to reattach at completion. Some of them will be fragile and will be attached with adhesives and may not survive. But save the ones you can. Put them in a safe place along with their fasteners before continuing with the job.

Get a couple of empty coffee cans or CoolWhip containers from your wife to keep fasteners in and keep them segregated. An issue often overlooked is bolt length. When you reassemble a machine there will be places where the 3/8"-16 X 1-1/4" bolts will work but the bolt that was there originally was 1-1/2" long. And when you come to the place where the 1-1/4" bolt came from you will find that the bolt you have is too long. So you put washers under the head and so forth and so on. Another thing about fasteners is that you should replace any fasteners which have battered or worn heads or screwdriver slots. It doesn't cost a lot and it is just one more detail that the craftsman will pay attention to.

Now let's go to the sandblasting pad. I have made two observations about sandblasting. Sandblasting should be done somewhere else — and by someone else. It is a dirty, messy, hot and irritating job. The fact remains though, that an hour of sandblasting will clean up machine parts which would otherwise require hours if not days of scraping, chipping and scrubbing and the end result will be a better job. Sometimes there is no way around the onerous jobs and this is one of those times. While we are on the subject of sandblasting let's explore it a little. Like every other shop procedure there are right ways and wrong ways to go about it.

First, what kind of compressed air supply do you have? Many catalogs

Photo 7. The big parts done and primed.

Rebuilding a (Your Machine Here)

Photo 8. And now most of the small parts are done.

which cater to the home craftsman will advertise sandblasting systems with promises of almost miraculous results if you buy their product. They will sell you the whole system. The pressure vessel, the hose and blasting nozzle, the protective hood and gloves and everything you need to sandblast your rusty lawn furniture like the pros do. But if you think you are going to hook this thing up to your 1-1/2HP, 7.3 CFPM air compressor and get the job done you are in for a letdown. Serious sandblasting requires serious compressed air. Something on the order of 75 CFPM at 80 PSI is what I would consider a minimum for pressure blasting. The big compressors required by sandblasting operations is one of the reasons why you will be charged high fees if you have it done by a professional but sometimes it will be worth the money it costs.

I do not consider myself to be any kind of an authority on the subject and I avoid it altogether unless there is just no good way of getting around it but sometimes I have to do it. I do my sandblasting outdoors in a rural community. I also have a sandblasting cabinet for the small parts. Check your local environmental regulations before you go to work with the nozzle. Make sure the wind is not blowing towards your wife's new Honda. Try to limit sandblasting operations to days when the humidity is low. If your jobs are of a size where you can use a sandblasting cabinet instead of working outside that is a good option. On the job we are working on here there were nine parts which were just too big to handle in the cabinet so I had to do them out on the pad.

And one more thing about sandblasting. Many parts will require additional attention before the final coat of paint. There will be sheet metal guards with dings and dents in them. There may be some cracks in frame parts or in castings or many other possible problems which will require your attention before you reassemble the machine. But regardless of any of that don't wait more than 24 hours before you paint the sand blasted parts with some sort of primer. Newly sand blasted parts will rust while you watch them. If you are going to fix the dents and dings that day OK but if the parts are going to wait any length of time at all then put some sort of protective finish on them.

Photo 9. Ready to begin reassembly. It takes a lot of parts to make a cutoff saw.

Some parts should not be sandblasted so you will need a good parts washing facility. Parts with machine finishes such as shafts which run in pivots or bearings are examples. The shaft in this machine about which the upper frame pivots and the screw for the vise are examples of parts which should not be sandblasted at all. The blade wheels were sandblasted in the cabinet but care should be taken to protect the bearings and/or bearing surfaces. As is the case in every evolution in my shop or in your shop or any other shop, common sense should always rule.

Divide your parts up into classes or categories. Make one pile of parts which will require sandblasting. Make another pile of parts which need to be wire brushed on the powered wire wheel. Another pile will need to be soaked in some sort of degreasing solution and then cleaned. Discard any fasteners which have damaged threads or rounded heads from being given the old "vise grip" treatment and replace them with new bolts or nuts.

It is a good idea to keep many of the subassemblies separated and complete them one at a time. If you are one of those people like the friend to whom I was referring earlier you will likely completely disassemble the machine and sandblast all of the parts which require it at the same time. I wish I could do that. But I find that I do better by taking subassemblies one at a time and finishing one up completely before beginning on another. In this example the two wheel assemblies demonstrate what I am talking about. The drive wheel with its gear box and planetary gears cannot be mistaken for the idler gear and its tension adjusting screws but — the frames which support them are very similar. If I were to disassemble both of them and do the required cleaning and finishing on all of the components there would be a strong possibility of getting the parts mixed up. And the results could be some minor misalignment of one or both wheels when I get finished with the job. So I finished one wheel assembly completely before I even started on the other. Blade guides were treated the same way and for the same reason.

After you finish all of the blasting and brushing and rubbing and polishing and painting it is time to start putting this thing

Photo 10. Beginning to look like a machine again.

Photo 11. Presto! A brand new machine. Well. Maybe not exactly "Presto."

back together. I did this job in spare time so I could not just set in and put everything back in one session. I have several four wheeled carts in my shop which get assigned to projects. I put the saw on one of them and that gives me the opportunity to wheel it out when I have the time, work on it, or maybe just sit and look at it, for a while and then roll it back out of the way.

You will usually reassemble in reverse order to the way it was disassembled. Begin with the base and whatever legs or supports which are involved. Another variation to this entire process is preparing the base and legs by sand blasting or scraping, painting them (or not) and set them up. Then address each added component one at a time by finishing or refinishing it, make it ready for use and the adding it back to the assembly. Usually the only person who really has to be satisfied with the process and its methods and eventual final results is you. The person doing the job. Even if you are doing the job for hire and rebuilding a machine for someone else, you will have to be satisfied before he is. I have never been one to subscribe entirely to the rule of the customer being always right because I have encountered some customers who — but let's not go there!

And that is pretty much all there is to it. You will of course want to put new grease in the new fittings you have installed. Then there is the chore of refilling the hydraulic feed cylinder and bleeding the air from the system. I left the coolant pump out of this saw because

Rebuilding a (Your Machine Here)

I can't conveniently use coolant in my shop and I didn't want the little gear pump to wear itself out by running dry. If you happen to buy this machine from me remind me and I will put it back on.

The photographs accompanying this chapter are all pretty self-explanatory and I have tried to provide captions which should clear up any confusion caused by my inability to describe a process in the text.

I hope this will give you some ideas and maybe prod you into getting that old DoAll or Hendey you acquired when your great uncle closed his machine shop out of the corner it has been sitting in for all these years and turning it into a show piece of useful machinery. As a matter of fact, I have a 16" DoAll already sitting on a four wheeled dolly which I may get going on. Now if I can figure out a way to avoid another session with the sand blasting hood!

Isn't it great how well we are looked after these days? Now that the lawyers have pretty much taken over it is just a matter of time until an end mill will come with the warning, "Caution! Cutting tools are sharp! Do not intentionally poke this tool into your eye!"

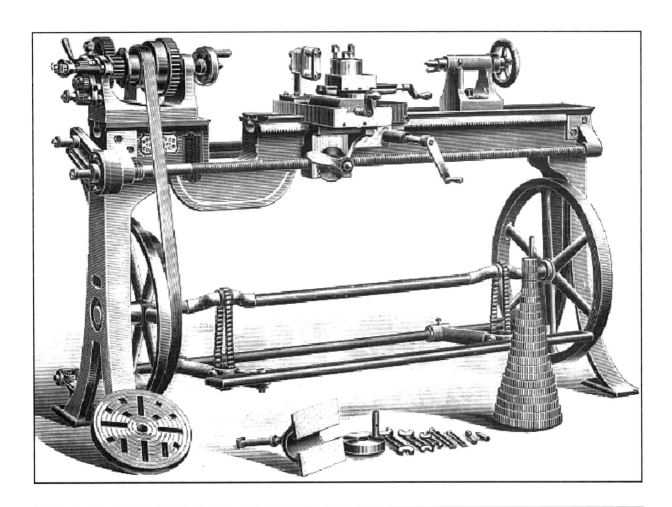

Chapter 4

Debunking the Myths of the Gap Bed Lathe

Myths of the Gap Bed Lathe

I was at an informal gathering of some fellow metal working enthusiasts several weeks ago and, as is usually the case at such an assembly, we were each describing some of the jobs we were either currently working on or had just finished in our shops. My contribution to the conversation was to relate the details of some large diameter parts I had been fabricating which had required the removal of the filler block from my gap bed lathe. The reactions I got from the collective group were surprising. There seemed to be a degree of awe or disbelief that I had actually removed that piece of cast iron and was able to put it back into place without destroying the future usefulness of my 17/24 inch engine lathe. 17/24 indicating, of course, the swing of the machine both with the gap in place and removed.

Since that meeting I have paid attention to other discussions about the same subject. I looked up some past references to the practice on the various Internet bulletin board discussion panels and the general impression seems to be along the same lines as that of the group where the subject first came to my attention. A lot of machinists, of many levels of experience, seem to be under the impression that, once removed and replaced, the filler block from the gap of a machine so equipped will never again be back as it was when originally assembled and that the machine will forever be incapable of producing accurate parts. Not so. It requires care and attention to several details but it is a pretty routine piece of work and should done routinely.

First, let's discuss for a moment the gap bed engine lathe. Probably more appropriately referred to as the gap lathe. The gap in these machines is a little like the spare tire in your pickup truck. It is a good thing to have when you need it. When you really need it you can't get along very well without it. But if you have to put your spare tire on every time you run down to the parts store then you should consider the cause and do something about it. I use the extra capability for swinging large diameter parts in my lathe perhaps twice or three times a year.

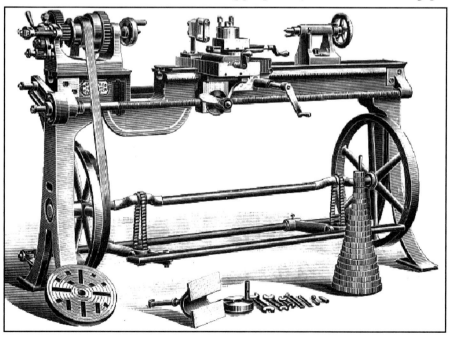

Photo 1. An antique, foot powered gap lathe from the 19th century.

I have made some tooling which makes this a little more practical, I will describe some of it here, and I have a customer which will bring me parts of that size at about that frequency. But if I were to find myself needing this capability every week or so and was having to spend a lot of time removing and replacing the filler block, I would probably be looking at either installing a larger lathe or re-examining the work load coming from my source of customers. I have seen machines in shops which were dedicated to specific jobs. In those cases the block was removed and stored and in at least one instance I know about, the block was accidentally disposed of. But for the most part, the removal of the filler block is something which will not be required often. When it is required it isn't that difficult to accomplish.

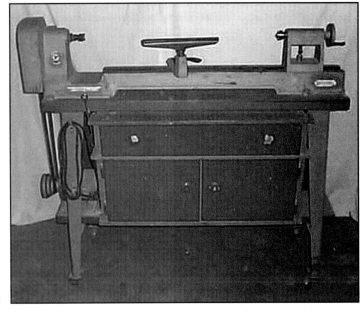

Photo 2. A gap lathe for making bowls or piano stools from wood.

Take a look here at some examples. Photo 1 is a picture taken from the internet of an early lathe telling us that the gap bed machine came along at a very early time in the development of the engine lathe. This was probably a pretty capable machine for its day. Besides having the capability to cut many different pitches of threads and to hold work of a fairly large diameter, the machine took the place of the treadmill in today's times! An eight hour day spent machining heat-treated 4140 on this machine would make you appreciate the soft chair in front of the TV when you got home, would it not?

Photo 3. A sliding gap lathe.

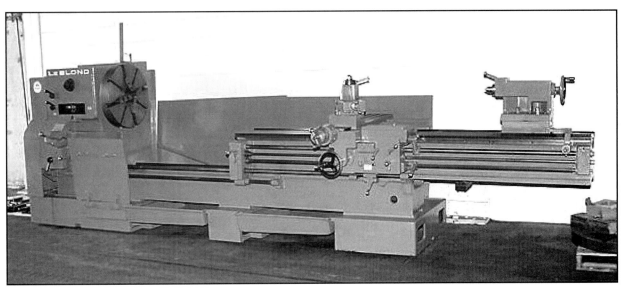

Myths of the Gap Bed Lathe

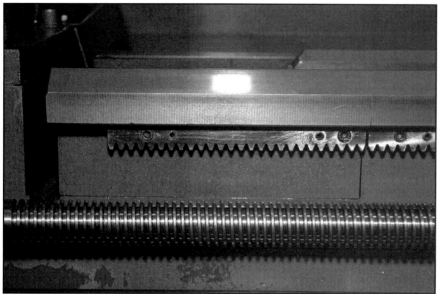

Photo 4. A view of the gap area of the lathe described here. A Kingston HJ-1700 capable of holding parts up to 24" in diameter in the gap.

Photo 5. The gap in my Kingston HJ-1700. Note the nuts for removing the tapered dowel pins.

Photo 2 is a picture of a wood lathe with gap capabilities. In this case there is no means of closing the gap but in most wood lathes, manual ones at least, that is not a requirement. This is just a further demonstration of what a little ingenuity can do in the design of the most simple machines.

The next photograph, Photo 3, is a picture of a different approach to the problem of swinging large diameter parts in engine lathes. I once had experience with a lathe of this type, also a LeBlond, in a power plant. Maintenance shops in power plants, maintenance shops in any kind of environment really, have to be as flexible as is possible when it comes to handling every job that comes in the door. This type of machine, known as a sliding gap lathe, is a good example of the versatile machines found in maintenance shops. On larger machines of this type, like the one shown here, the sliding upper bed is powered by a motor driven screw and on smaller machines there is usually a crank at the tail stock end of the machine. There is an advantage to using this type of gap machine. If the machine is equipped with good way wipers and reasonable care is taken when moving the slide it makes it easier to maintain the accuracy of the machine when changing from one configuration to the other.

Photo 4 is a picture of the section of the bed which can be removed from my gap lathe. A Kingston HJ-1700 purchased new in 1997. And since it is the only gap lathe around here right now it is the machine we will use as an example in the following job example.

The best way I can think of to demonstrate all of the steps involved in gap lathe machining is to follow an actual job through from start to finish. I have the advantage here of using only the photographs and descriptions where the results were good. I will carefully avoid showing you the facets of the job which I screwed up. (I can't believe I said that!) We won't be discussing the stripped thread on the locating taper pin and the subsequent process of drilling out the pin. The mashed finger, the lost 10mm hex wrench and the piece of material which had to be re-ordered because of poor communications are also parts of this job which will serve no useful purpose in being talked about here. We will describe the process of building a special tool holder and the methods for holding work pieces of this size, however, and anything else which may shed light on using a gap bed engine lathe.

I would show you a drawing of the part but it is a proprietary design used by a good customer so I better not go there. I will provide a general description. The finished part was a shallow dish made from a piece of C1119 steel plate, 23-1/2" in diameter and 2" thick. I would probably hesitate to begin machining a piece of plate this size if it were made from heat treated 4140 or another of the tougher steels. There are limits to what can be done on small machines.

Photo 6. Top, taper pins removed and (6) bolts loosened.

Photo 7. Above, gap section removed and in its temoporary storage place.

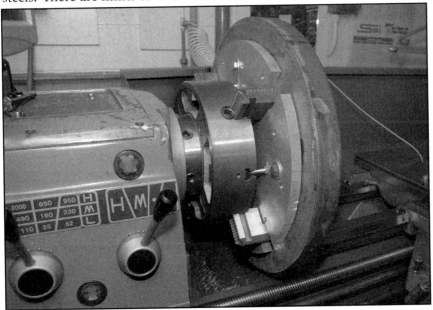

Photo 8. The back side of the part. Bolted to an aluminum plate, which is in turn chucked in the 4-jaw chuck.

Myths of the Gap Bed Lathe

Photo 9. The part viewed from the front. Note the extended tool holder in place in preparation for facing the part.

Photo 10. A different approach to holding large diameter work.

Begin by removing the gap block from the machine. The two taper pins shown in Photo 5 have threaded ends on them and, most of the time, (see my note in an earlier paragraph about things we do not need to discuss here!) can be withdrawn by using a nut and a couple of washers. Then remove the bolts, 6 of them on this machine, as shown in Photo 6. Photo 7 shows the gap section removed and stored on a piece of wood I keep for that purpose

Photo 11. An aluminum plate chucked in the specially constructed chuck jaws.

at the end of the lathe. Most jobs which are large enough in diameter to require the removal of the gap section will be short and the stored section will not be in the way by placing it here.

Photos 8 and 9 are of the part set up in the lathe preparatory to machining. In Photo 8 you can see the aluminum plate I have used for several parts in this series of fixtures. There are no firm rules here. You have to assess the requirements of the job and set it up accordingly. In this case I needed to machine the outside diameter and the entire face of the part. Drilling and tapping holes in the back side was not going to interfere with the eventual use of the part but that will not always be the case. An alternate option, used on a previous job, is shown in Photos 10 and 11. Photo 10 is of a set of specially fabricated chuck jaws which are bolted to my 3-jaw chuck and which are used to hold the part shown in Photo 11. Photo 11 is of a different job, by the way. There are many other

Photo 12. Two special made tooling bars I have made for machining large diameter parts. They may also serve as heavy boring bars.

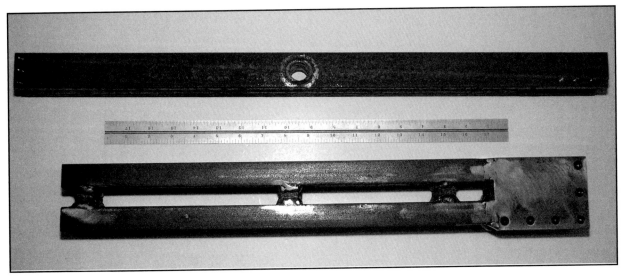

Photo 13. Tapping the setscrew holes in the head of the bar.

Photo 14. Set up for welding.

alternates and each job will have its own peculiarities and problems. Remember that the machine does not limit the methods nor the results. The operator is the limiting factor. And you are the operator.

I am going to digress here for a moment to talk about some tooling I have made and which I use whenever possible when machining in the gap. One of the most important things, in my opinion, about gap machining is the avoidance of allowing the saddle wings to hang over the gap where the underside is exposed to swarf and debris. Sometimes it cannot be helped but when you can do something about it ---- do it. The extended tool holders I will describe here is a good solution to this problem. With these tool holders, I can perform almost all of the machining operations required yet still not have the saddle lose part of its support nor have its bottom side contaminated with a bunch of crap. There will be some loss of rigidity but compromising is what gap machining is all about.

The tool bars pictured in Photo12 are two that I have made for use on my 17" lathe and Photos 13, 14 and 15 are pictures taken of the process of making one of them. I am not including dimensioned drawings for these bars as they will vary so widely depending upon the type and size of machine they will be used on; the type and size of job they will be used on; and upon the availiblity of materials from which they will be constructed. They may also, as many of my tools do, reflect the mood you are in when you make them.

Photo 15. Another view of the setup for welding. The angle plate used here is an old one I use just for welding setups.

However many variations of extended tool holders there may be, however, they will all have two things in common. They will all be made rigid enough to support the expected loads at the tool tip and they must all be made so that the tool tip is presented at the centerline of the lathe axis. So the first thing you must do is measure the height of the centerline from the top of the compound. See Photo 16. Be sure to consider the size and configuration of the tool you will be using and then be careful in doing your arithmetic. Look back at Photo 9 and you can see the finished bar and cutting tool in place to begin facing the part.

After the job is finished comes the part which seems to cause the most concern among users of these machines. Putting the gap block back into its former position in the bed of the lathe. It isn't rocket science but it is important that care be taken in doing this. First and most important, clean the chips and swarf and other mess from the bed of the machine. Brush it all out clean and then wipe it out with a clean shop rag and a mild solvent. The solvent serves mostly to insure that the way lube or cutting oils have been removed. The oils themselves will do no harm but they will hold tiny chips and particles and that is what we are after. Get this area really clean! After cleaning all of the foreign materials out, run your hands over the bed of the lathe and make sure there are no burrs. Remember back when you were

I never figured this out. The people who spend the most time trying to find the cheapest tool are the ones who spend the most time complaining because it is cheap!

Photo 16. Measuring the height of the machine's axis centerline from the top of the compound.

Facts can sometimes be a real nuisance when telling a story and should be ignored as much as possible.

about half finished with the part and you dropped your adjustable wrench while changing the tool angle? Where that wrench hit may have caused a slight dent in the lathe bed and where there is a dent there is a corresponding high place. That is what you are after.

Using a clean, fine grit stone, stone the area completely. There should be no reason for the insert to have suffered any trauma but make sure of it as well. After stoning any areas you suspect may have dings or burrs on them, clean the entire area again using solvent and a lint-free shop rag. As a last measure I will put a fine film of a light lubricant on the area.

Now, carefully, set the insert back in place. Put all the bolts in hand tight and then tighten them as you would a cylinder head. Reinstall the tapered dowels at this time after cleaning the holes out with something like a bore brush for a .45 cal. gun barrel. Tap the pins lightly into place at first and then drive them home after the bolts are snugged up the first time around. Tighten the bolts at opposite corners using slight pressure on the wrench and then go around them again and tighten them all down really snug.

And that is all there is to it. Granted that, if you have the ability to measure to an accuracy of .00001" you will find that the block is not in the same place as it was when you removed it. But if you use reasonable care in removing the block, handle it carefully while it is out, and be thorough and careful in cleaning it and reinstalling it, it will go back into place well within the accuracy limits you or your lathe are capable of producing. Otherwise you will just have to buy a bigger lathe. Which ain't a bad idea, either!

Chapter 5
The Turret/Ram Milling Machine

Turret/Ram Milling Machine

I have a copy of TMEH, the *Tool and Manufacturing Engineer's Handbook, Volume 1, Machining,* which frequently serves me well as a beginning place to look for information on many machine shop operations and practices. My copy is the Fourth Edition and was copy righted in 1983 by the Society of Manufacturing Engineers. I looked in it in preparation for writing this chapter and was surprised to see that, of nearly 80 pages of information on milling machines, there was only one paragraph about, and one photograph of, the turret/ram type of milling machine. There is a lot of competition for attention in milling machine types and apparently the big guys have an edge. But the turret/ram machine is the predominant machine found in the basements and garages and back yard shops of the fraternity of home shop hobbyists and machinists. They will also be found in tool rooms and in the tooling corners of machine shops in almost every manufacturing facility in the country. Sometimes they serve only as glorified small drill presses but they serve as well in that capacity as they do everywhere else. I don't have any data about how many of their smaller cousins, the table top machines known mostly as "mill/drill" machines, there are in comparison but it is probably a significant number.

The turret/ram machine, likely universally known in our fraternity as the Bridgeport type of machine, is so named because it consists of a movable ram mounted atop a turret swivel base which is, in turn, mounted to the top of a column. Most of the ones available today from our friends across the oceans, especially from the far side of the Pacific Ocean, will have as minimum features:

- Either a variable speed or step-cone pulley on the spindle and a back gear
- Powered quill feed
- 9" X 42" table

Other optional features might include:

- Power feed on one or both horizontal axes
- Coolant pump and reservoir
- 2- or sometimes 3-axis digital readouts
- Larger work tables

There are still many of the namesake Bridgeports around with the 'round ram' over arm and 36" tables which date from the origin of the company to whom the genre of machines owes its name. Their origins date to sometime back in the 1930's. However, the imports have generally taken over the market for the home or hobby shop.

I am basically satisfied with my life and the way I have led it. But if I could make one change I would try to get finished with serious welding before I reached the age of 50. Since then it seems that if I get in a position where I can see the weld I can't breathe. And vice versa.

Manufacturers and distributors of these machines use a lot of phrases like, "massive castings" in their advertising. Apparently some of them have never seen a massive casting but that is beside the point.

Many of these imported machines are quite serviceable. They wouldn't stand up long in a busy shop where several different operators might be using them three shifts a day but, when used by the person whose money paid for it, and when reasonable care is taken of them, they can do accurate work for quite a long period of time. The accuracy and versatility of any machine tool is, as we all know, a function of the operator and not of the machine.

It is not the intent of this chapter, however, to proclaim the serviceability or lack thereof of any particular machine brand or type or even continent of origin. What I am going to try to do here is to answer some questions that novices might have and which are answered neither in the manual which came with their machines nor in *Machinery's Handbook*. Some more experienced people may find an answer or two here as well but they will all be like me and will say, "I already knew that." We will talk some about a few of the many accessories available for these machines but I will be limited to speaking with any authority only about the accessories I own with my machines. You may consider that any accessory I do not own is one that I desire and I know I am not alone in that regard.

For purposes of our discussion here I am going to divide work done on these ubiquitous little machines into two camps, holding the work piece so that the tool can cut it and holding the tool so that it can cut the work piece. I am going to open this discussion by talking some about a few of the methods and devices used for holding the job and presenting it to the cutter. There are many options.

Now I don't want you to start writing letters to me about the methods I leave out here, either in work holding, tool holding or available options. The ingenuity of the machinist is limitless. In all of my years of working in and around machine shops, and there have been more of them than I sometimes like to think about, I still from time to time will see an approach to a job that I never saw before. And some of them are pretty simple yet effective. I'll give you an example. I saw a photo in a copy of *Model Engineering* magazine of an arbor support which mounts to the column of the machine and which enables the operator to do a more efficient job of machining with arbor mounted cutters. It was simple to make, easy to use, highly effective in the job that it does and just something I had never thought of making. We will explore it in more depth in Chapter 18. You probably already knew about it!

Use the tool to do the job for which it was designed. Good advice. But sometimes that adjustable wrench in your back pocket does make a great hammer.

Photo 15 is a picture of the device on my milling machine.

Bolting Work Directly To The Table

Excluding setups where the work piece must be rotated or moved in other than X or Y axes during the cutting process, this is the option which should almost always be considered first. Mounting and clamping the work directly on the table eliminates all of the joints or possible sources of unwanted movement between the table and the point of cutting. See Photos 1, 2 and 3 for examples of how and how not to set work up using T-nuts, studs and flange nuts.

Photo 1: How NOT to do it.

- The heel or rear support of the clamp is lower than the part being machined. This causes a force pushing the work piece away from the clamping pressure.

- The supporting heel is unsuitable. The use of the Woodruff cutter and spur gear is an obvious exaggeration but I have seen worse.

- The nut does not have a washer under it. A hardened flange nut with built in washer should be used but at the very least there should be a washer under the clamping nut.

- The distance from the stud to the work piece is greater than the distance from the stud to the heel. When this condition exists the holding force on the part is reduced. The stud should be as close to the part as is possible without interfering with the cutter or the collet.

- Only one clamp is used. When circumstances and work piece size allows there should always be at least two clamps.

- The cutting tool here is mounted in a Jacobs chuck rather than a proper collet. This is more of a tool holding issue than work holding but it doesn't hurt to be reminded.

Photo 2: How TO do it

- The heel is raised so that it is higher than the part being clamped. The toe of the clamp pushes straight down on the work piece instead of pushing away.

- The supporting heel is solid and substantial. The 123 blocks at the bottom of the serrated block spread the pressure out so that the heel block is not directly over

When there is more than one way to do a job, nine times out of ten the method which requires the most effort is the best way.

the T-slot in the table.

- There is a flange nut on the clamp which reduces the likelihood of the nut deforming and becoming loose because of the slot in the clamp.

- The stud is located as close to the part as it can be thereby increasing the mechanical advantage by moving the fulcrum. The pressure applied by tightening the flange nut is multiplied.

- There are two clamps on the part.

- The cutting tool is mounted in a proper tool holder.

Photo 3: Another example of doing it right.

- This was a simple machining job of cutting a 16mm X 16mm slot across the face of an Instron testing machine bed plate. Notice that the clamping bolts are offset from the X axis so as to allow the cutter to traverse all the way across the part unimpeded.

These are just examples. There will always be compromises and the trick is to know when to compromise and in what area. Some things to be considered in any setup are:

- Never set up a part so that you run the risk of machining through the part and damaging the table. There are no excuses for this, period!

- Consider the job before you set it up. Make

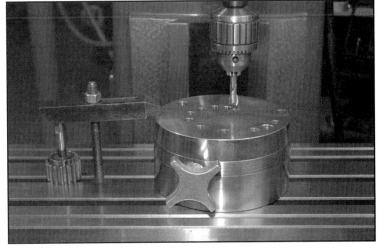

Photo 1: How NOT to do it.

Photo 2: How TO do it.

Photo 3: Another example of doing it right.

allowances for cutting paths as shown in Photo 3. A little thought before turning the spindle on can sometimes save having to set the job up again.

- If there is more than one machining operation to be considered set the part up so as to be able to accomplish all operations, or at least as many as possible, in one setup. It may involve moving one clamp while the other clamps on the job are left in place. But there should always be at least two clamps engaged.

- Repeatability of operations should also be considered. If there are multiple parts to be machined the setup should include stops or fixturing to allow for changing parts without running the risk of losing accuracy.

It will be said over and over, and it is always true, that no two jobs are alike. And no two setups are alike. But all jobs and all setups have this in common; if the tool pressure exceeds the work holding pressure something is going to give. And when that happens something is going to take place you would just as soon not see. It can be a ruined part, a broken cutting tool, a damaged machine or fixture or, worst of all, an injury to the operator. There are short cuts in the shop which can save you time but taking a short cut in setting up a job which can or will result in any of these disasters doesn't make sense. Don't do it!

Vises

I have many vises. My friends tell me that I also have many vices

Series 1 Photos

Series 1, Photo A — A drill press angle vise. This is a heavy vise I acquired in a trade. It is useful for machining angles or for drilling holes at other than 90 degrees to the table surface. For the record, the scars and the drill marks on the vise were already there when I acquired it.

Turret/Ram Milling Machine

Series 1, Photo B — A 3" Kurt swivel vise. The vise I use the most on my milling machine. I keep this vise and a pair of tie down bolts handy on my tool caddy at all times. I will sometimes, but not often, remove the swivel base from the vise.

Series 1, Photo C — The bottom side of my 3" Kurt. The keys are installed in the bottom of the swivel base so that the jaws are aligned with the X-axis of the machine every time I bolt it down. I have keys similar to these on every work holding accessory I use regularly. If I have to swivel the vise to produce an angle, then of course it must be re-indicated after that job so that the jaws are once again parallel to the machine table.

Series 1, Photos D, E, and F — This is a vise from an old Cincinnati No.2 Tool and Cutter grinder. This can be a really useful accessory when you are required to machine odd, compound angles. The vise and its 90 degree mounting brackets may be positioned so as to present almost any angle to the cutting tool. Here is a good example of a compromise. Every time you add a joint or adjustable mating surface to a setup, you decrease the rigidity of the setup. You will pay the price by having to take lighter cuts at reduced feedrates but sometimes the sacrifice has to be made to accommodate a job.

Building Shop

Series 1, Photo G — Another small swivel vise.

Series 1, Photo H — A 6" Kurt vise and a homemade mini-vise. I have the swivel base for the Kurt but keep it unattached because of the reduced Z-axis operating room on my machine. When I need it I know where it is.

but at my age a vise is more useful to me than a vice. The milling machine vise is probably the most used accessory for holding work on the milling machine table and is second in rigidity only to bolting directly to the table. The series of photographs, see Series One on the accompanying pages, represent a few of the vises I own and use. The Kurt ™ vises and others of the design which provides downward pressure to the movable jaw are relatively recent, if thirty years ago can be considered recent, and are good designs. The advent of CNC machining centers has required that work holding be made more uniform and we manual machinists are beneficiaries of that.

All of these vises and others not mentioned here have one thing in common. They are useful work holding devices when the work required involves machining in X, Y, Z or W axes. If there is a need to cut curved profiles, or if cutting multi-faceted parts such as square or hex parts, or gears, then we must go the next level of work holding.

Photo 4 — A simple, 5-C indexing mechanism.

Indexing/Rotating Mechanisms

This discussion will not extend to all of the actual methods for using these indexing mechanisms, but will rather be a listing of them and will hopefully provide a description of each along with the times when their use should be considered. A comprehensive text covering all of the methods for selecting and using indexing fixtures would be great subject for an entire book. Maybe next time.

There are, broadly speaking, two classes of indexing fixtures widely used on the turret/ram milling machine. One provides a method for positioning the work so that subsequent sides may be machined in turn, as in machining the six sides of a hexagonal part or the four sides of a square. The part is clamped in position, machined on one face, then unclamped and rotated to the next position, clamped once again, and the process repeated all the way around the part. The other type of fixture allows the work to be machined as it is rotated. Here are a few examples of what we are talking about.

Photo 4 is a picture of a simple, 5-C indexing mechanism. In Photograph Series 2, by the way, you can see some of the variations of the different types of collets and other work or tool holding accessories being discussed here. This indexing fixture is called a 5-C because of the configuration of the collet it accepts. This work

Turret/Ram Milling Machine

Photo 5. Indexing head and slitting saw arbor.

Photo 6. Universal indexing head and gear cutter.

64 Building Shop

Photo 7. Closer view of the gear cutter seen in Photo 6. Note the set screw on the lathe dog.

holding accessory is useful for machining each side of a job in turn but it cannot be used very handily for machining while the part is being rotated. (I learned a long time ago not to say that something cannot be done because somebody will do it and write me a letter. I only say that it cannot handily be done. There are better ways.) This is a useful device in that it may be quickly mounted on the milling machine table and does not require a lot of setup time. There are keys on the bottom of this fixture similar to the ones shown on the bottom of the Kurt vise in Photo C. This fixture may be mounted in the vertical position but provision must be made to access the collet tightening knob.

Photo 8. An 8" rotary table with a centering plug in the Morse center hole.

If more complex parts, a 19 tooth gear perhaps, or a larger work piece than may be conveniently mounted in a collet, need to be machined the dividing head is where you will be heading. Photo 5 is a picture taken

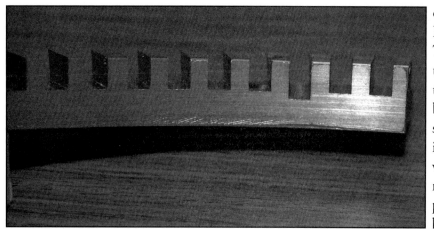

Photo 9. An "oops." What happens when you don't keep an eye on the cutter in the collet.

Photo 10. The accessory lug on the back of the ram. Found on most light milling machines.

of a simple dividing head I will use occasionally. This head approaches the upper limit of the size of tooling fixtures which may be conveniently used on a small milling machine. But it does the job many times when other methods will not. This photo is obviously posed with the slitting saw but it shows the method and the use of a three jaw chuck. Other chucks may be mounted on the head and I do have a four jaw which will be used on this accessory occasionally.

Photos 6 and 7 are of another dividing head which is even larger and more complex than the one in Photo 5. It also is a little large for use on our small home shop milling machines but it works and that is what counts. Notice that in this application we are using a tail stock, sometimes called a foot stock, to support the work piece between centers. This is not necessary, a chuck may be installed on this dividing head the same as the one in the preceding discussion, but it does provide a different approach. Make the process fit the job, whatever the job! In this example the work piece is mounted on an expanding mandrel which in turn is driven by a bent tailed lathe dog. Notice the setscrew necessary to keep backlash out of the setup. This is a useful method for cutting gears and a gear cutter is shown on the machine arbor. This dividing head is an old Kearney Trecker accessory and it could probably tell us some stories. It has the capability, when used in conjunction with a "real" milling machine to rotate the work piece as it feeds through the cutting planes in order to produce a helix. This feature is difficult to provide on the small milling machine because the feed mechanisms generally operate from a separate driving power source but when this is what

Turret/Ram Milling Machine

Photo 11. Above, using the rotary table in conjunction with the slotting head.

Photo 12. Left, "poor man's DRO." Using the dial indicator as a DRO.

Turret/Ram Milling Machine

Photo 13. An oval profile machined using the Volstro Rotary Milling Machine Head.

I am told that the second faculty we lose as we get older is our memory. I just can't remember what the first one is.

you need then this is what you need ---- no doubt about it. On the small milling machines which we are discussing here it can only serve as a versatile and useful indexing mechanism. With both this indexing head and the one shown in Photo 5 it is possible to machine parts while they are being rotated.

Probably the most widely used indexing mechanism to be found in small shops is the rotary table. Its usefulness is limited, as is the usefulness of all of our tools, only by the imagination and resourcefulness of the operator. By mounting chucks or vises or other work holders of various descriptions on the device, and by being able to divide 360 by the number of divisions we are required to produce, there are really hardly any limitations to what we can do with this tool. It has the ability to drive a part through a cut to produce circular profiles or it serves simply as an indexer. It may be used in conjunction with the vertical slotter or with the rotary spindle. Some of the rotary tables I have seen, but do not have an example to show you, have an indexing plate built in so as to provide indexing capabilities without having to use fractions of a degree. Photo 8 is a picture of the rotary table which receives the most use in my shop. It has a number three Morse taper through the bore and I have made many centering arbors from worn out Morse drills,

Turret/Ram Milling Machine

Photo 14. Left, the Volstro installed on the Seiki milling machine.

Photo 15. Below, a variation of an outboard arbor support.

Building Shop

Turret/Ram Milling Machine

Series 2 Photos

Series 2, Photo A — 4 configurations of 5-C collets. Left to right, a ½" round, a ½" hex, a ½" square and a pot chuck. 5-C collets will accept work pieces up to 1-1/16" in diameter and are also made to accept square, hexagonal and other shapes as long as their outline can be inscribed in a circle which does not exceed 1-1/16". Larger parts can be held in the 5-C pot chucks. They may be machined to accept a larger diameter part than a standard collet. The plug seen in the center of this one allows the fixture to be machined while under compression.

Series 2, Photo B — An R-8 collet. Used almost exclusively as a tool holding collet in the spindles of small milling machines. Usually available in sizes from 3/16" to 7/8".

Series 2, Photo C — A Weldon type end mill holder. The chief disadvantage of this holder is that it will allow you to install a cutter in your spindle which may overload your machine. The example shown here will hold a tool with a 1-1/4" shank which is a pretty big load for a 2HP milling machine.

Series 2, Photo E — Quill Master. This attachment is useful sometimes, I have used it, but only in limited applications. You may find it more useful than I ever have. A great way to break small end mills if you are not careful.

Series 2, Photo F — Tapmatic tapping head. For the times when you have repeated holes to thread. Mostly a production tool. You will break fewer taps with one of these than you will by hand.

Series 2, Photo G — Criterion boring head. Practically indispensible in my shop. I have even used it to machine the OD of a raised boss on a part by running it CCW but you must be sure that it doesn't unscrew itself from its arbor while doing this.

Series 2, Photo D — An R-8 to Morse adaptor. Very useful if you have counter bores with Morse shanks or if you want to hold a Morse drill. Notice the slot for using a knock out drift.

Series 3 Photos

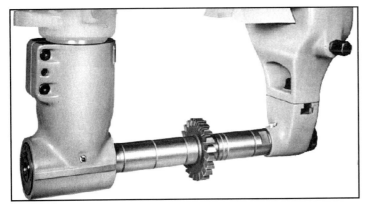

Series 3, Photo A — A right angle attachment with support yoke. This allows the vertical milling machine to be used as a horizontal machine. Particularly for operations like cutting gears.

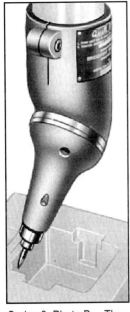

Series 3, Photo B — The Quill Master we have already discussed. I still think it is more of a novelty than a useful tool but when you need it, etc.

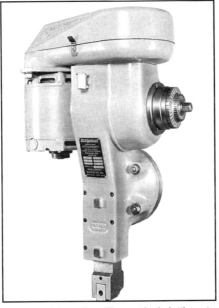

Series 3, Photo C — The vertical slotting attachment. The stroke has 6 speed settings, from 70 through 420 strokes per minute, and is adjustable from 1/8" through 4" in 1/8" increments.

Series 3, Photo D — A cherrying head. A unique piece of equipment. This provides the cutting tool with an oscillating movement through either convex or concave arcs for machining die cavities. This is an example of how ingenious mechanical movements were employed before CNC machines came along. This is one of those pieces I cannot justify owning but I want one nevertheless.

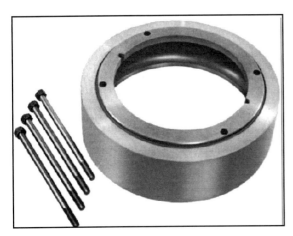

Series 3, Photo E — Riser blocks. Available in several heights this adds to the depth, or height, of a work piece which may be mounted under the quill.

one of which may be seen in the picture. It is not the intent of this chapter to discuss all of the methods for using rotary tables, nor any of the indexing mechanisms, but it does need to be said that this is an accessory that you will eventually need if you do any mill work at all.

A discussion about indexing/rotating mechanisms would not be complete without describing the Volstro Rotary Milling Head. I also like to take advantage of any opportunity I can to boast a little about owning one of these accessories for my milling machines. Volstro machine accessories include many of the other machine accessories discussed later on which were made, or at least sold, under Bridgeport's name. But to my knowledge, which we all know is flawed, Bridgeport never marketed a device exactly like this one. Photo 13 is a picture of an oval profile machined using the Volstro. The radius of each end is 1-1/2" and the distance between the centers of the radii is also 1-1/2". This profile was machined with a ¼" roughing mill using the DRO in combination with the rotary milling head. You can readily see where a profile of this configuration would be difficult to machine accurately using any other method short of a CNC machine. Photo 14 shows the accessory installed on the quill of the Seiki milling machine with the timing belt in place. The timing belt provides the opportunity to drive the circular motion of the unit using the quill feed mechanism of the milling machine. Just don't engage the feed clutch while using it.

Many of the milling machine accessories used in our small shops over the country to produce anything other than straight lines and square corners are like the Volstro Rotary Milling Head. They are ingenious and wonderful mechanical devices and they are as obsolete as the cross cut saw. CNC controls and ball screw axis movements and servo motors have relegated most if not all of these tooling accessories to the back shelves of the tool room if not to the scrap heap. But they make wonderful toys and provide a capability to our home shops which otherwise would not be available to us.

The Other End Of The Job

Now we will go to the other end of the job. Holding the cutting tool and the many ways of doing that. Here again there are many options and they all have their place. The rules are clear — don't hold an end mill in a Jacobs chuck — don't climb mill with a heavy cut — use only center cutting end mills when plunging — and on and on. And the rules are good rules and should be adhered to whenever we can. But — what if you need to make a cut in a place where there is no practical way to hold the tool besides a Jacobs chuck? The rules are

Your neighbor will rave to you about the wonderful new titanium golf club he got for Christmas but doesn't understand why you had rather talk about your new zero to 12" dial caliper.

like all rules and can and should be bent from time to time. Don't misunderstand me here. I am not an advocate of taking what may turn out to be expensive shortcuts. If your reason for substituting less than the optimum tool to do a job is that you are lazy, and I know a little bit about lazy, by the way, then don't do it. But knowing when the hard and fast rules can be suspended in order to get the job done, and do it safely and without undue risk of a negative outcome, is not a bad thing. There is no way for me to say here what should or should not be done in every case. Just use your experience, your intelligence, your common sense and the tools you have at hand and go to it!

In the type of milling machine we are discussing here most machining you will be doing will either be axial, as in drilling, reaming or tapping, or it will be radial, as in cutting keyways or milling slots or other surfaces. When cutting axially the side pressures on the cutting tool are practically non-existent. The pressures are in line with the spindle and generate thrust only in Z or in W axes. This, by the way, is the chief reason that I am against using one of the variations of XY positioning tables on a drill press. Drill press quills, even big ones, are designed to cut axially and do not usually have the capability to withstand radial loading. When cutting radially, feeding the work piece along while the cutting tool cuts on its side, the cutting pressures are pushing against the quill of the machine and for this reason, you should do all of this type of machining with the quill retracted as far as is convenient. Sometimes it has to hang out and there will be nothing you can do about it but remember the compromises. The less rigidity you can provide the tool the smaller the cuts you will be able to take and therefore your productivity will go down. And productivity counts, even in a hobby shop! The preferred method for holding cutters in this application is what is known, at least by me, as a Weldon type end mill holder. See Photo C of the pictures in Series 2. This provides a positive lock and will prevent, or at least minimize, the effect shown in Photo 9. Yep! I screwed up! I know that is a surprise to all of you who think I never make mistakes but sadly, here is the proof. The series of slots shown in this picture were machined in a piece of A-2 tool steel. The slots were radial, that is they were a series of slots radiating from a common center and were being cut in a part which was to be a chuck jaw for a large machining center for a customer. The 5mm end mill making the cuts was a two flute end mill held in an R-8 collet. The cutting pressures applied caused the helix of the end mill to pull itself out of the collet. I knew better. I just thought I could get away with it.

We tend to remember the most the mistakes which cost us the most and this one was expensive! I had about 3 hours already invested

I have an edge finder in my tool box. I have a center finder and an angle finder in my tool box as well. Now if I only had a chuck key finder and a TV remote finder I would be fixed.

in this part, not to mention about fifty dollars worth of A-2 steel, and now it is useful only for a paper weight. And I already have plenty of them! A good trick, by the way, to avoid this is to spray DyKem on the shank of the cutting tool where it enters the collet and then pay attention to it. If it shows a tendency to pull out you will immediately see a band of shiny tool shank at the collet. I didn't do that and paid the price. A Weldon end mill holder would have prevented the problem. One caution should be mentioned about Weldon type holders. Just because you can devise a method for holding a cutting tool in your spindle does not necessarily mean you can successfully machine a part with it! We have all seen large drill bits, sometimes 1" or bigger, machined down to fit into a 3/8" drill chuck. The same thing applies. Keep in mind that "boring with a big auger" usually takes a big machine. Don't try to put a 5 HP tool in a 2HP spindle and expect much in the way of good results!

Other tool holders often used on the small milling machine will include a Morse to R-8 adaptor for when you have to drive a drill bit or a counter bore, the adjustable boring head, the tapping head and sometimes, the Quill Master. These are all shown in the photographs and I include them because I have them in my shop. I have used them all at one time or another. I will use the boring head frequently, the Morse adaptor fairly often, the tapping head three times a year or so and the Quill Master once in a blue moon. But I have used it.

More Stuff

The fully equipped turret/ram milling machine has many faces. If you have ever wondered what the protruding lug on the back end of the ram casting on your milling machine is for, I am about to tell you. It is 4-1/2" in diameter and 3" high and it has a 7/8" hole through it and it is used to mount various machine accessories. Photo 10 shows it better than I can tell it. The most common, or at least the one I hear about the most, is the slotting head, or vertical shaper. That is the accessory I have mounted on my Lagun and I find it is frequently useful. Used in conjunction with the rotary table or with one or another of the indexing mechanisms, it is possible to machine internal splines or other internal or external profiles which do not lend themselves readily to other machining methods. Photo 11 is a picture copied from Chapter 8 of my first book, *Randolph's Shop*, and provides a good example of using the slotting attachment.

I have taken a liberty by copying from the Internet several representations of tooling accessories available for the turret/ram milling machine. These were all offered by Bridgeport at one time or another. And some may still be available from that source but most of these items are now largely available on eBay or other used machine suppliers.

And Still More Stuff

I feel the same way about people who claim to have "fully tooled" milling machines as I do about people who say they have a "full set of drill bits." I do not dispute the claim outright but I have serious doubts. A fully tooled milling machine of any configuration involves one heck of a lot of tools. Here are a few of the things we haven't talked about in this chapter and I still will be leaving something out.

Digital Read Outs — Both my machines are equipped with DROs and I would hate to give them up. Photo 12 is a picture of what I call a "poor man's DRO" and I used this system for a long time. I still use a dial indicator to monitor movement in the W axis of the machine. My DRO's, in addition to keeping track of table movement, in either inches or millimeters, have the capability of calculating bolt patterns. Three axis systems are available but I do not find the third axis to be as useful for me. Depends upon what you are doing.

Power feed systems — Another useful accessory but if you can only afford either a DRO or a power feed go for the DRO. I do have power feed on the X axis of both my machines and I do use it frequently. Power feed is also available for Y and W axes on these machines but here again, I doubt they would be really useful to me.

Flood coolant — One of my machines has flood coolant and one does not. I have used it but for my application, and I think for most hobby applications, the disadvantages of using flood coolant outweigh the advantages. If ever a coolant is developed which will not evaporate nor deteriorate nor get contaminated with machine lubricants nor begin to stink because of any or all of those problems, then I will use it on every job. Until then I will get along without it. On high production, high speed CNC machines it is a way of life.

Power draw bar — I have used these and do not find them to be as helpful as the idea suggests. If you are too short to reach the top of the machine or if you change tools really frequently, you may disagree but for me it is something I can do without.

In these pages I have talked about milling machines and a few of the accessories available for use with them. I have likely just scratched the surface. If you look in any of the many tooling catalogs you will find a thousand different tools and devices made to serve two purposes.

The stated purpose is to make your life easier in your shop but the real purpose, I sometimes think, is to separate you from your money. I don't mean to unjustly criticize but some of the "indispensable" tools being offered are just barely short of ridiculous. I am too smart to mention any one of them here, by the way. But you have to decide if it is useful to you or not. I have mentioned a couple of times here that I find the Quill Master accessory by Bridgeport to be marginally useful. But I know that there are many machinists who have used this tool to a great advantage and who would disagree with me. It is a diverse world and we are a diverse bunch of people. And home shop and hobby machinists are probably as diverse as any group to be found. Can you imagine preferring to spend time in a basement shop, making something which, when complete, will only be valuable to you and then only because you are the one who made it? Who ever heard of such a thing? I wouldn't have it any other way.

---

*Wanna know how to make the bottom of a bell hole weld look as good as the top?
Screw up the top!*

Chapter 6
It's Never Too Broke to Fix

It's Never Too Broke to Fix

It's never too broke to fix.

That statement will probably cause some of you to pause and to think about broken things you have seen and you will say, "I bet I can come up with something that is too broke to fix." And that is true. Some things are truly in the condition I refer to as Class III Broken. If it is Class III Broken you better know where you can either find a replacement or be prepared in some way to get along without it. But many times the condition of "broken" an item or an assembly is in, relates directly to the amount of effort we are prepared to put into making the repairs. And if we are prepared to really go the extra mile in making repairs then usually, it isn't too broke to fix.

This job was brought to me by a friend who seems to have a knack for finding jobs which most people would turn down at first sight and declare to be past economical repair. But he is the best machinery rebuild guy I know. I have observed in the past and am still of the opinion today, that I would prefer to own a machine which was rebuilt by Gary than to own the same machine brand new. He is that good. I don't know why he continues to waste his time working on old boats when he could be working full time on rebuilding machines.

Photo 1. Above, the broken casting. Notice the name plate on the side. this should be removed before beginning the repair.

Photo 2. Right, another view of the casting. Somebody thought we might not notice that the part is broken!

Photos 1 and 2 are two views of the job before we begin. The casting is, or at this stage of the game, was, the adjusting ramp for the outfeed table from a Northfield jointer. The notation you see written on the side of the casting in Photo 2 indicating, "broken", probably

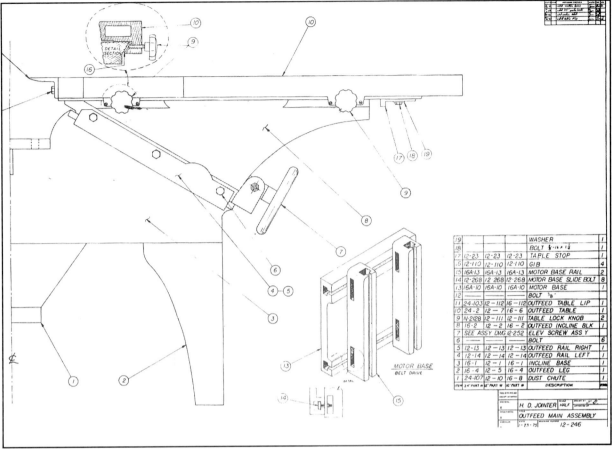

Drawing A. A drawing of the original part. You should always provide yourself with as much information regarding the part you are repairing as you can find.

should have said, "broken and lost" because that was what made the job more difficult. But not impossible. The broken section of casting was either discarded in some housecleaning effort or used as a boat anchor or possibly, it was broken into too many pieces at the time of the accident for the owner to think it worthwhile to keep the parts. Whatever happened, we are left with a partial casting. Fortunately, we have resources. The parts list for this machine was available and it included a drawing of the broken casting which showed enough detail to be able to come up with a pretty good pattern for use in the repair. Drawing A is a picture of the part we need. The assembly drawing did not include dimensions but enough detail was present to allow us to scale the part. Other details of the missing piece of casting were determined by comparing to the infeed side of the machine which was still intact. There is one other thing to take notice of in Photo 1. The name plate on the side of the casting should be removed before going further. Clamping the part on the welding table and sliding it around to weld it will cause damage to the plate and the only real way to avoid that is to remove it and then replace it after everything else is completed.

After making the decisions about what was required we went to

You will never enjoy a week's vacation quite as much as you enjoy the week of anticipation leading up to the vacation.

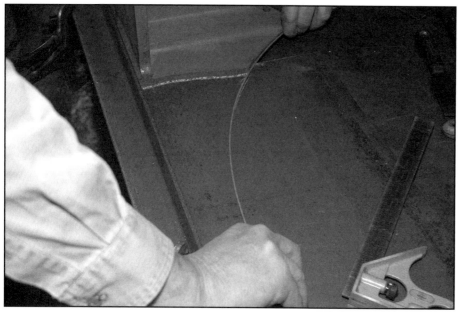

Photo 3. Laying out the profile with a piece of flexible wire.

work making the replacement parts. In Photos 3 and 4 you see an adaptation of the old French curve method of laying out a curved profile through two or more points. This can be useful when the curve is not an exact arc with a known center point.

Generally the profile serves as a blending agent rather than having to fit something exactly. In this case the two sides of the casting were broken along different lines so we had to lay out a left and a right part. 1/2" steel plate was a trifle thicker than the thickness of the original castings so it was chosen as the repair material and the offset caused by the difference in material thickness was put to the inside so that it would not be noticeable. These parts were burned out and can be seen clamped into position in Photo 5. One of the cross members was also burned from 1/2" plate and the remaining members were sawed from barstock of the appropriate cross sections. All of the component parts of the repair assembly were then clamped into position for welding as may be seen in Photo 6.

Photo 4. Ready to burn out with the oxy-acetylene torch.

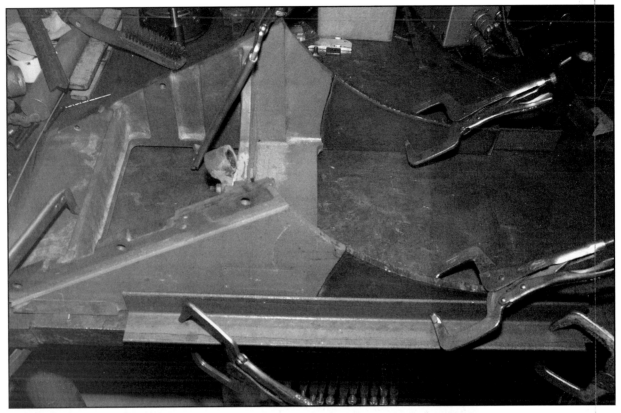

Photo 5. Above, side plates burned out and clamped into position.

Photo 6. Left, crossmembers sawed to length and ready to weld.

Photo 7 is another view of the clamped up assembly and is included to point out an important feature. Notice that the horizontal weld joint has been beveled to allow for more complete penetration of the weld. Examine the assembly for areas like this where the welds will be ground flush with the adjoining material and be sure not to create a condition where the weld will be

It's Never Too Broke to Fix

Photo 7. Above, end view of the ready to weld assembly. Notice the weld prep ground on the near end. Also if you look closely you can see the difference in the thickness of the casting and the material chosen to make the repair.

Photo 8. Right, welding completed. The twisted piece is just a temporary brace.

weakened. This area will be the end of the jointer table when complete and you do not want a crack to be showing here.

Photo 8 is a picture taken of the completed weldment. The twisted bar in the foreground (the twist is obviously not a necessary feature — it was just a handy piece of material) is tacked into place to help keep the assembly from becoming distorted while welding and will be removed. 100% welding is not necessary here for strength. The original casting, however, would have included fillets at all of the corners and the intent here is to make the repaired part look as if it had never been broken. Another way of accomplishing the same effect, and it is simply a matter of choice, is to skip weld the joints, making sure there is enough weld to provide adequate strength and then add the fillets with one of the fiber glass products used in automotive body repair shops.

Now for what some may consider to be the hard part. Welding the completed repair assembly to what is left of the original casting. I am a great believer in adding studs to a cast iron weld wherever it is possible and on this job it was possible. Set the casting up in the drill press as shown in Photo 9 and drill and tap for the studs. If you do not have a drill press big enough or if the geometry of the part makes machine drilling difficult then drill and tap by hand. A

Part of being a skilled workman is knowing what NOT to try to do.

Photo 9. Preparing to drill and tap for the studs.

It's Never Too Broke to Fix

Photo 10. Right, reinforcing studs in place and sawed to length.

Photo 11. Below, close up view of the weld joint ready to weld.

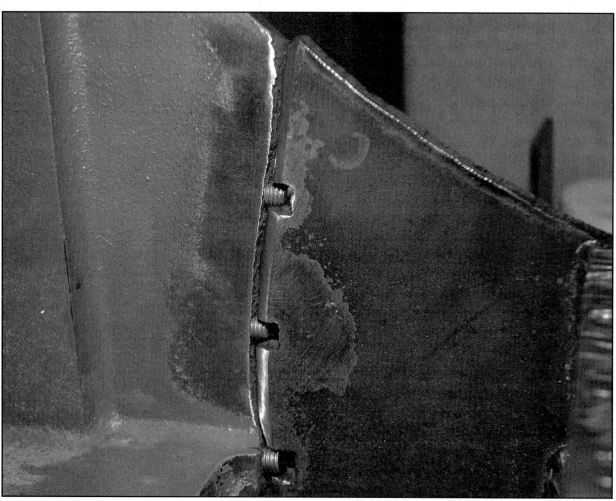

84   Building Shop

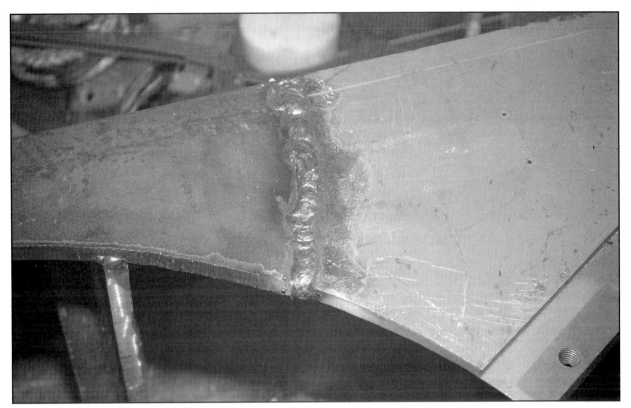

Photo 12. The completed weld.

couple of studs embedded in the cast iron will make cracks much less likely to start in the weld and will add greatly to the strength of the weld so it is worth some extra effort to put them in. Don't try to put studs in that are too large for the part. Trying to put a 1/2" diameter stud into a cast iron web that is only a few thousandths of an inch over 1/2" thick, for example, is not a good idea. I used 3/8"-16 studs on this job. 5/16" diameter would also have worked and even 1/4" would have been better than leaving them out altogether. Photo 10 shows the studs in place while Photo 11 is of the repair assembly notched and ready to weld. Notice the bevel ground for a weld prep on the steel part. Grind the bevel on both sides of the weld.

The usual precautions for welding cast iron all apply here. Use an electrode in which you have confidence. I use a Eutectic product called XUPER 2233 which has always given me excellent results. I welded this using DC reverse polarity at about 110 amps. As is always the case with cast iron, preheating to about 500 degrees F and maintaining that temperature throughout the process is a good idea but is probably not absolutely necessary here. Your own experience will tell you what to do with different jobs and they are all different. You can see the finished weld in Photos 12 and 13. Be sure that you fill in all undercut areas and that the corners are full enough to be finished and blended without leaving holes in the weld.

Beware the customer who wants to provide the material for his job. "This is a piece of really good steel" is a great endorsement but it doesn't tell you much about what to expect when you machine it, weld it or put it in the furnace.

Grind the welds and fillets to blend with the original casting. Photo 14 shows the start of that process. This is where the appearance of the job will be most effected. Don't overgrind and make ditches along the welds. If there are dimples and undercuts, and there will almost inevitably be some, use body putty to fill them in and finish the job by blending the radii. Photos 15 through 18 were taken after the welding repairs were completed and the machine was being reassembled. A useful trick here is to add a handful of casting sand to the paint when applying the finish color. This is of course purely cosmetic but will further contribute to the illusion of having an original appearance. Notice particularly in Photo 17 how the repaired section looks as if it were never broken — which is the real sign of a professional repair. This is the sort of job which will keep people coming to your shop as long as you are willing to stay there for them. And that ain't a bad thing!

Photo 13. Another view of the completed weld showing both sides.

Photo 14. Beginning the process of grinding the welds..

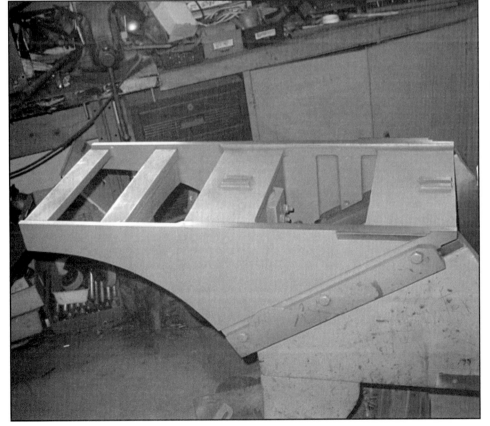

Photo 15. Above, reassembled jointer bed with repaired casting in place.

Photo 16. Left, repaired casting. Note welded area.

Photo 17. Above, closeup of repaired casting. Looks like the real thing!

Photo 18. Right, another view of repaired casting.

A political contradiction: Sometimes people who are dumb enough to run for an office are too dumb to do the job if they get elected.

Chapter 7

Ornamental Iron — Again!

Ornamental Iron

Wrought iron, it seems, just will not go away. Not that the product should be done away with, just the term itself. Wrought iron is a term used almost universally by people less informed than us iron workers to describe ornamental ironwork. By far the most common material used to fabricate ornamental railings, signs, furniture or other pieces of hand forged products is plain old, hot rolled, mild steel. Wrought iron is another product entirely. But I don't suppose it matters in the long run what people call it. There is a demand for it and it doesn't make sense to get into an argument with a potential customer who might be willing to spend a large sum of money for something you can produce in your shop. You will have plenty of other opportunities to lose paying customers, believe me.

The job I am going to do here for this chapter is just one example of this type of work. There is no limit I am aware of to the number of jobs available in most communities in this country involving some simple forging, a little welding and some imagination. This particular project was done in memory of a prominent citizen in my community who was also, coincidentally, a good friend of mine. He was involved

Photo 1. Sign made by John Bulgin, about 1960.

Ornamental Iron

Photo 2. Ornamental sign built for cemetery.

in arranging for the sign described here to be built and installed at the old cemetery owned by a local church when he died suddenly in a fall at his home. Let that be a lesson to all of us, by the way!

Photo 1 is a picture of a sign made by my father in about 1960. The First Presbyterian Church in my home town is a building of historical significance, having been built in 1854 and which survives today. It has been restored in recent years to its original configuration including the pot bellied stove although it probably is not the same stove. But I am getting away from the subject here. I know you all have a vital interest in the history of my home town and we will take that up at another time. Meanwhile, back at the forge…

The sign shown in Photo 2 is the sign I built for the cemetery and which we will use as an example for this chapter. Keep in mind that this chapter is more about methods than it is about product. But do feel free to copy this design if you have a similar place for a memorial sign or even just a sign at your driveway. I have made many signs similar to this one over the years including a couple for my own shop. Photos 3 and 4 are of the signs I have at my shop and there has never yet been a bill collector or a tax assessor who has not been able to find his way here.

Some self-proclaimed "tool and die" experts behave as if they have been smelling Dykem too long.

Building Shop

Ornamental Iron

Photos 3. Sign at Randolph's Shop.

Begin the construction by fabricating the post. This one is made from 3" square tubing with 3/16" wall thickness. A lighter wall thickness will work as well since strength is not so much a consideration as proportion. Don't use something so light that you have difficulty welding it and keep in mind also that, depending upon the climate where you are, rust and corrosion may be a problem. Rust will usually attack the post where it enters the ground so the material thickness

Ornamental Iron

I don't mind loaning my tools. All is ask is for a signed deed to a couple acres of real estate in the event the tool is not returned.

Photo 4. Sign at Randolph's Shop.

Building Shop

Ornamental Iron

Photo 5. Layout of point at top of post.

Photos 6, right, and 7, below. Progress of job to this point.

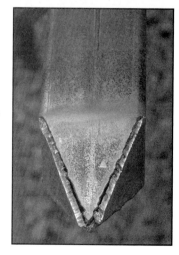

I am getting so forgetful that I have to make a list of things I need to do during the day. Now if I could just remember to look at the list.

may be important here. In Photo 5 you see the layout of the point at the top of the post. Measure down from the top of the post a distance equal to the width of the post and draw the lines as shown. Burn out the corners and form the point. You will probably have to use some heat from the torch at the bend lines. Photos 6 and 7 show the progress of the job to this point. Weld up the corners and grind them smooth as seen in Photo 8.

Now for the fun part. Laying out the

94 Building Shop

ornamental elements of the sign. Or of the picture frame. Or of the decorative structure you are building to adorn the entrance to the shrine where you keep your most prized Harley-Davidson. Or whatever. I wish I had the ability to describe what I mean here when we are talking about 'proportion'. The result you are looking for must look – you know – proportionate and – nice. And we all have different notions of what looks nice. The most skilled artisan or worker of materials of any type is not effective if his or her designs do not have an appealing look to them and I don't really know how to describe it. Usually a good piece of advice in these designs is to keep it simple. Scrolls and geometric repetitions add to

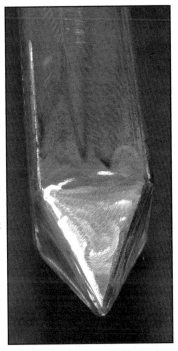

Photo 8. Left, corners welded up and ground smooth.

Photo 9. Below, laying out the ornamental elements of the project.

Ornamental Iron

a design but there will be a point where additional scrolls will start to detract from the overall picture and you will have to know where that point is. I wish I could offer more help than that, but I can't.

I can tell you that, once you have decided on the finished design, you should lay it out in full scale. Draw the elements out as I have

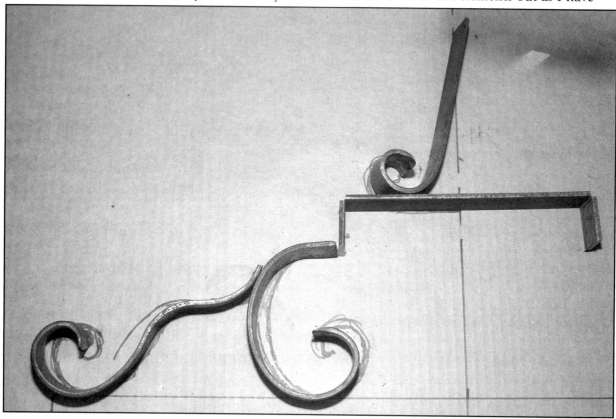

Photo 10 above. Using drawing to obtain correct shape of scrolls.

Photo 11, right. All finished elements before welding for final assembly.

96   Building Shop

done here on a piece of a corrugated pasteboard box or piece of plywood. It will simplify the job if you draw only one half of the required elements where that is possible. Photo 9 is of one side of the top ornaments and in Photo 10 you can see how this is used to obtain the correct shape of the scrolls. Photo 11 is a picture of all of the finished elements of the job before welding them together in the final assembly.

Photo 12. Making a finished end on the scroll material.

I want to make one comment here on the scrolls themselves and this is mostly personal preference. I have seen many jobs done where it is obvious that the scrolls were bent from sawed material and the ends are left as they were when sawed off. Sometimes you can see the saw burr still on the end and that is not merely personal preference – that is just plain laziness! I prefer to make a finished end on the scroll material as seen in Photo 12. It doesn't take a lot of time and it adds a finished look to the job that will be seen by the people who appreciate craftsmanship. In Photo 13 you can see the first welding of the scrolls in progress and the 1/4" thick spacers on the welding table are there to keep everything aligned and in the same plane.

Photo 13. First welding of the scrolls in progress.

Photo 14 was taken after the project was primed and before

Ornamental Iron

Photo 14. Project has been primed but not yet finished.

painting with the finish coat. I use a light colored primer on jobs of this nature for a reason. It is difficult at best to paint a job with a lot of scrollwork. You have to paint it at least 4 times on each side in order to get complete coverage and the contrasting color of the light primer to the darker metal helps to insure you don't miss anything. The same is true then when you apply the finish coat. Black is the traditional color of iron work, although there are many other choices, and the contrast of the black finish coat to the gray primer serves the same purpose.

I had the best of both worlds on this job. I did the ironwork and somebody else provided the hole. Two volunteers from the church congregation dug the hole and mixed the concrete to install the sign. All I had to do was go take the picture after the sign was in place. The sign itself was done by a local sign painter.

This was a satisfying job. I marked the finished job by stamping my friend's name on it along with the date. The group who shared in the project received a nice note of appreciation from our friend's widow and the sign is now there as a reminder of who he was. But we still miss him at our weekly breakfast meetings.

Chapter 8
If You Don't Want to Take the Job — Just Say So

If You Don't Want to Take the Job — Just Say So

A customer brought a job to me which is a perfect example of situations frequently encountered in the small machine shop. That is to say, it had all sorts of built-in excuses for not taking the job. Most small shops are begging for work so they will usually take whatever comes in the door, but some people are whipped before they start. They don't have the ingenuity, or the ambition, or the whatever-it-takes to overcome the difficulties presented by the "out of the ordinary" job that comes their way from time to time. The job used here is, as I have already said, a perfect example of what I am talking about.

The job is shown in the drawing and in the accompanying photographs. It is a 17" diameter piece of aluminum, 3" thick, which requires a concave profile to be machined on one of the faces. The profile is of a specified radius, in this case 40". Here are the objections which might be heard:

1. I can only swing 17" maximum diameter in my lathe. I don't have room for chuck jaws to grip the part.

2. My biggest chuck will only accept a maximum diameter of about 16". I have no way of holding the part to machine it.

3. My lathe is a conventional engine lathe. I have no good method of machining contours like the 40" radius called for.

4. The 15-1/2" diameter bolt pattern on the back side is too big for my milling machine. I only have 15" of

If you work with your hands and you are proud of what you do — mark it. Place some sort of mark on the work which you will be able to identify in later years. You won't regret it — unless you are making something illegal.

Photo 1. Extra large capacity chuck jaws. These things can be dangerous if you don't pay strict attention to what you are doing.

If You Don't Want to Take the Job — Just Say So

travel on the Y axis of my machine and I would need about 20" of travel to accurately place those holes.

Now if the customer who brought the job is one who has a poor record of paying their bills or if you just don't like him or her for some reason, these are all good excuses for not taking the job. Any one of them can be a show stopper if you really want to stop the show. Unfortunately, there are many machinists for whom these

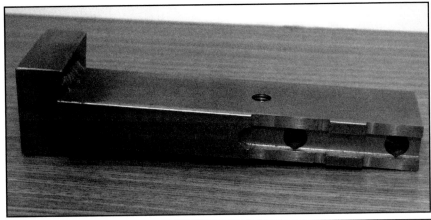

Photo 2. Left, details of chuck jaw construction.

Drawing of aluminum duck's nest part.

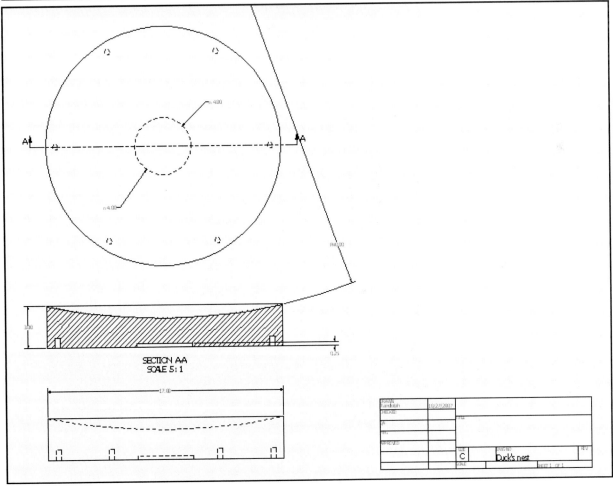

Building Shop

Photo 3. Positioning the material in the machine. Some sort of material handling assistance is welcome in a job of this size.

Photo 4. The left end of the guiding rod. Notice that the reader head for the X-axis of the DRO had to be removed for this job.

reasons, one or more of them, would be enough for them to not go any further. But as I have said many times and will continue to say as long as I am up and walking around in a machine shop, the limitations of the capabilities of a shop are dictated by the limitations of the man operating the machine. Not by the machine itself. I can cite too many examples of situations where the ingenuity of the machinist overcame the limiting features of the machine whether it be the size, the speeds available or maybe even the color of the paint. Let's talk about the reasons given above.

Excuse number one is a pretty easy one to overcome. We are working with a gap bed engine lathe here so we simply remove the gap and now that objection goes away. There is an in depth discussion about gap machining in Chapter 4.

Excuse number two requires a little more thought. Photos number 1 and 2 are of the chuck jaws I made specifically for this job. The jaws are welded up and then bored in place to accommodate the part. Use heavy material for this application. I used 1" X 2" CRS for making the jaws because anything much lighter would deflect and not provide the positive gripping force needed for these parts. Photo number 3 is a picture of actually chucking the part in the machine. The overhead crane I built to serve my machining area comes in handy here.

Excuse number three is where we start getting innovative with the job: Machining the 40" radius on the face of the part. This is an old trick which I learned many years ago and which still comes in handy today as you can see here. But there are some conditions attached. The concept is that you force the cross slide of the machine to move in an arc by jamming a rod between the tail stock of the machine and the cross slide.

Photos 4 and 5 show the dimples drilled into the tail stock and the cross slide of the machine. I drilled these with a hand drill and they only have to be deep enough to make sure the guiding rod does not slip out of place. The two studs seen protruding from the side of the cross slide in Photo 4 are the mounting posts for

Photo 5. Tail stock end of the guiding rod. The dimple in the base of the tail stock was drilled just deep enough to insure that the point wouldn't slip out.

the X-axis scale from my DRO. I had to temporarily remove it from the lathe for this job. The way this works is that the rod forces the saddle to move in the arc described by the length of the rod. But — the required radius is 40" and the rod I used is only 28" long. The correct procedure is as follows:

First do a little arithmetic. Get out your *Machinery's Handbook* and find the formula for calculating the height of a circular segment for an arc chord 17" in length on a 40" radius. This turned out to be .915". Set the compound of your lathe to travel parallel to the spindle axis and touch the tip of the tool to the center of the part. Now move the cross slide towards the tail stock far enough to clear the job and advance the compound feed by .915". Measure the distance from the tool tip to the back, or right, side of the cross slide. In this job that came to be almost exactly 12". Subtract that figure from the required radius and you come up with, or at least I did, a length of 28". That is the length of rod you will need to produce the 40" radius. Grind or machine an appropriate point on each end of the rod and you are in business. The rod, by the way should be heavy enough to resist any tendency to bow in compression. The loads are not great but a ¼" diameter rod would probably bend. I used a piece of ¾" CRS for this application. Photo 6 shows the rod in place

Photo 6. An overhead view showing the rod in place between the cross slide and the tail stock.

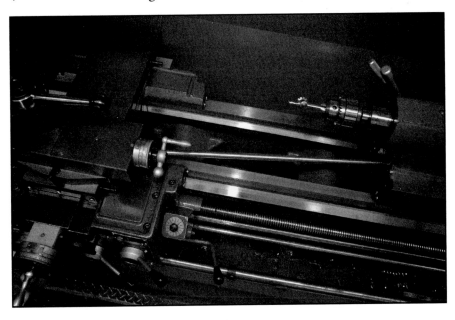

If You Don't Want to Take the Job — Just Say So

Photo 7. Taking the first cut.

Photo 8. Below, I fed the compound in about .125" per cut.

Photo 9. Nearing the final cut. Remember to hold pressure against the guiding rod with the carriage hand wheel when backing out for the next cut.

between the cross slide and the tail stock which is clamped firmly to the bed of the lathe.

Because we are using a 28" length of rod between the tail stock and the cross slide, and the balance of the 40" is a straight line which moves parallel to the machine axis, the resulting 40" radius will not be exactly true. If an exact profile is required some method of placing the end of the rod closer to the tool tip will have to be devised. In the job being done here the level of accuracy was not that demanding. Just be aware of the geometry of the movements produced by this method. Photos 7 through 9 are of various stages of the cutting process and photograph 10 is of the finished profile.

Excuse number 4 moves to the milling machine. When attempting to drill and tap a series of holes on a part this big using a small turret ram machine you will run into interference. If you set the part up so that you can reach the holes closest to the column then you will run out of space before you can reach the holes on the opposite side. There are a couple of methods for getting around that. The most simple is shown here but it depends upon the accuracy required.

First locate the center of the part as nearly as you can using

104 Building Shop

Photo 10. The finished profile. Facing big parts like this will always result in surface finish differences over the face. CNC machines with constant surface speed capabilities are about the only sure way around this.

conventional methods such as the center finder of your tri square. Place a center punch mark at the exact center of the part. Position the ram of the milling machine to approximately 5-1/2" as shown in Photo 11 and, using an appropriate center, position the spindle directly over the center of the part as shown. See Photo 12. Calculate your hole pattern from this position and drill and tap the holes closest to the column of the machine. Next, without removing the part from the table, re-position the ram to the 12-1/2" location as shown in Photo 13; re-center the part as before; recalculate the hole pattern and drill and tap the remaining holes. Remember that the positioning figures given here are for the specific job being done here. Your requirements will change as your jobs change and they will always change. If you are more comfortable with indicating the part between the movements of the ram then that is the way to go. The method shown here can, if you are careful with re-positioning the ram and the spindle, provide an accuracy of less than .010" which is adequate for most jobs.

This is just one job. There are countless variations of jobs you will be presented with and each one will have its own peculiarities and requirements. Just remember what I said. The limitations are always a function of the operator and not of the machines. And when you need an excuse to turn down a job you can always find one.

I wonder what the guy who named the EZ-out did for a living.

If You Don't Want to Take the Job — Just Say So

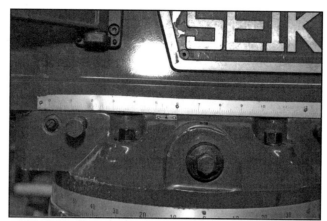

Photo 11. Positioning the ram of the milling machine to drill and tap the holes closest to the column of the machine.

Photo 12. Locating the spindle at the center point of the part.

Photo 13. Below, the ram now moved out for drilling and tapping the holes on the outer side.

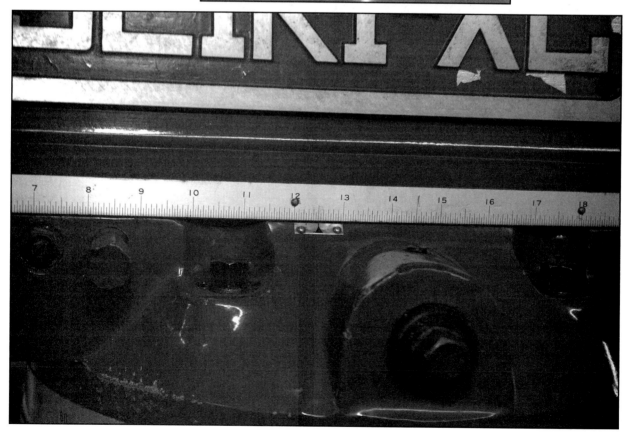

106   Building Shop

Chapter 9

Building a Gun Safe

Building a Gun Safe

The title of this chapter, Building A Gun Safe, may be misleading. We are building a gun safe, after all, only if we keep guns in it when completed. If the builder chooses to keep bullion in it, it then becomes a bullion safe. Or a watermelon safe. You get the idea here. By using these plans as a starting point and by modifying them according to your needs you can build a safe, or a vault, to keep business or family records, jewelry and valuables, or, depending on the size of your shop and other resources, you can build a place to keep that restored 1964 Ford Mustang. OK. That may be stretching the point a bit but the idea is that here is a starting point for a project which can be modified to suit your own needs. I am including a materials list for this project but it applies only to a safe of this size and configuration. You will have to change materials according to your needs.

The cabinet of this safe is 56-1/2" high, 39-1/2" wide and 30" deep.

Photo 1. The vault I built in 1984. I used different hinges and handle on this safe but there are not any really significant changes. I didn't have access to the welding equipment back then that I do now.

It was designed and built to store guns, both long guns as in shotguns and rifles, and handguns. Our society today has made it necessary to lock up our sporting guns along with guns kept on hand for self defense or the defense of our homes. And it should go without saying, but it needs to be said anyway, that any gun should be kept out of the reach of children. Period. This safe is a good way to do that. Photos 1 and 2 are of two safes I have built. Photo 1 is a safe I built in 1984 and which I still have and use today. Photo 2 is the safe shown here in the photographs, drawings and explanations I hope to present you with. I am likely guilty of owning a safe in which I have invested maybe $500 in materials and probably $2,000 worth of labor, assuming my labor is worth anything at all, into a project that will be used to lock up and protect about $1,000 worth of valuables. But that would not be unusual in my case. My projects are often worth less when finished than the total of the expense I went to in creating them. I have no problem living with that.

Now for the inevitable disclaimers. This safe is lined with 1" of ceramic fiber insulation, which is advertised to provide thermal protection at up to 2300° F. I read the literature and I bought the material and I installed it in both my safe and in this one which was built for a friend. But the only way I can guarantee that the treasures inside the safe will survive a fire is to test it by burning down my shop and my dedication to the craft does not extend to that point. The same proviso applies to protection from theft. If you build a safe according to these plans, whether you alter them or not, and somebody has the resources and the determination to break into it, you will have my heartfelt sympathy but that is far as I can go. I will, however, have to say that if anyone does break into this safe, he is a remarkably capable thief. Either that or a good man with an oxy-acetylene torch!

One other set of precautions I will point out. This project when finished weighed just a smidge under half a ton and some of the components, besides being heavy, have sharp edges and corners. There are many opportunities in a job this size to strain muscles, cut or mash fingers or otherwise cause you pain and/or discomfort. Some sort of lifting device such as the ones discussed in my first book, *Randolph's Shop,* is, if not absolutely necessary, then at the very least highly recommended.

Photo 2. The safe described here in this chapter. Waiting here to be loaded on the trailer.

The drawings included here are pretty basic. They will provide details of the individual components, particularly of the locking mechanism, but the photographs are where you will get the most information about how to go about building this project.

Now, if all of the provisos and dos and don'ts are taken care of, let's get started. As noted previously, the materials list given here is for a specific vault. If this works for you then that is well and good but if you have different needs, or different shop facilities, or a different

Materials List For Safe — 56-1/2" High — 39-1/2" Wide — 30" Deep

Material For Front Frame, Door And Locking Mechanism:

1/4" Plate:
- (1) piece 35-1/2" X 52-1/2" — the door itself
- (6) pieces either sawn or burned to 8" diameter circles — locking discs

1-1/2" Square tubing — 1/4" wall thickness — frame for door:
- (2) pieces 51-1/2" long
- (2) pieces 39-1/2" long
- (4) pieces 4" long ---- locking plunger guides

2" Square tubing — 1/4" wall thickness — front cabinet frame:
- (2) pieces 56-1/2" long
- (2) pieces 39-1/2" long

3" diameter CRS — Hand wheel hub: 4" piece

2" diameter CRS — locking disc hubs and axles:
- A 12" piece will make these parts

1" diameter CRS — hinges and handle spokes:
- 36" piece will do this

5/8" diameter CRS — locking pins:
- (4) pieces 6" long

5/8" square HRS — locking links:
- (2) pieces 20" long
- (4) pieces 10" long

1/4" X 1" strap — locking pin links:
- (8) pieces 1-1/2" long

1/2" X 1-1/2" strap — hinges:
- 18" total

(3) 1/2" diameter steel balls — hinge bearings

(3) 1/16" NPT grease fittings

(2) bronze bushings:
- 1" bore X 1-1/4" OD

(2) snap rings to fit 1" shaft

(12) 1/4" X 1-1/2" long Grade 8 bolts with fiber lock nuts — locking link attachment points

Maybe a handful of miscellaneous hardware: 1/4" key, 1/4"-20 screws, etc.

Material For Box:

3/16" Plate
- (1) piece 55" X 38" — back
- (2) pieces 27-1/2" X 38" — top and bottom
- (2) pieces 27-1/2" X 55" — sides

1-1/2" X 1-1/2" X 3/16" structural angle
- (4) pieces 28" long
- (2) pieces 56" long
- (2) pieces 39" long
- (4) casters — MSC catalog page 3171 — MSC number 66027863
- 4" X 2" Medium/heavy duty

5/8" - 11 threaded rod — adjusting legs
- (4) pieces 8" long with (2) nuts each

Paneling and carpet — inside lining

Insulation — KaoWool ceramic fiber blanket — McMaster/Carr #93315K64 — 8#/cu. ft

1/4" - 20 X 1-1/2" long bolts — 50 — w/nuts and large area washers

Sargent and Greenleaf combination lock — M/N 6730

I would be surprised if you do not make many material substitutions for the items listed here. Anything goes when planning a project of this type.

Photo 3. A cross section of what the door frame and the frame of the door look like. Notice that the locking pin extends through the inner frame and well into the outer frame. The wall thickness of the inner frame should be the same as that of the outer. I just didn't have a piece of the proper material for the photograph.

sized place to park it, or if you just want to be different for your own reasons, then modify the list to accommodate your project.

Building the Front Frame, the Door and the Locking Mechanism

Begin construction by grinding a bevel on the outer edges of the door. Photo 3 shows a cross section of the edge of the door assembly. The bevel is not absolutely necessary but it will make it more difficult for a would-be safe cracker to get a start with his pry bar. Then lay the door on your work surface and start cutting the frame parts. Saw the 2" square tubing to length with 45 degree ends. Next saw the 1-1/2" tubing to length and place it in place in relation to the outer frame. Tack weld a couple of 1/2" spacers between the 2" tubing and the 1-1/2" tubing on each side of the frame to keep the parts aligned while drilling the locking holes. See Photo 4. Take each side in turn to the drill press and drill the locking holes from the inside. Drill through both sides of the 1-1/2" tubing and through one side only of the 2" tubing. When you are finished drilling put a short piece of 5/8" diameter rod through the holes and then clamp the parts back onto the door front. Be sure to remove the 1/2" spacers

Building a Gun Safe

Parts needed to build the hand wheel, the hinges, the locking discs and their mounting hardware

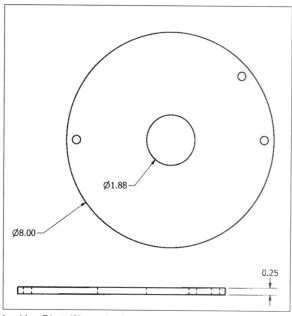

Locking Disc: (6) required.

Hand wheel hub: (1) required.

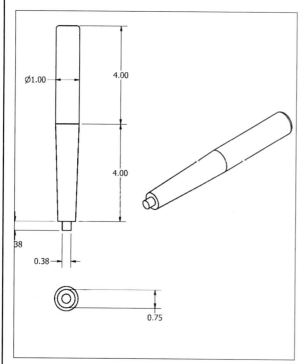

Hand Wheel Spoke: (3) required.

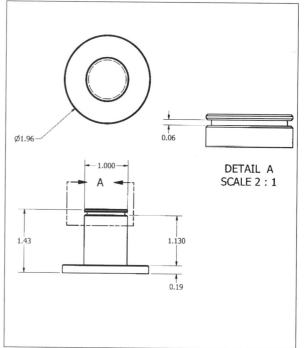

Locking Disc Axle: (2) required.

Building a Gun Safe

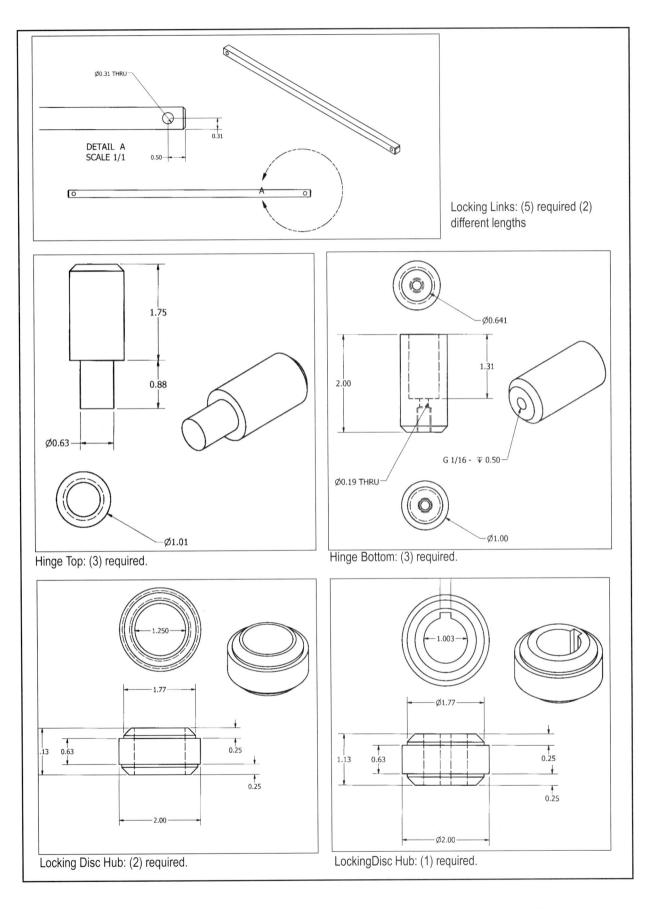

Building a Gun Safe

Photo 4. The door frames held relative to each other by using a 1/2" spacer tacked between them. They must be at each corner and must be removed before welding up the door assembly.

Photo 5. A picture of the corner before welding. 45 degree joints are a little more trouble to make but they do result in a neater looking weldment.

before you begin welding up the door assembly. When you have finished sawing and drilling you should have a clamped up assembly with corners as shown in Photo 5. Photo 6 is of the complete door frame assembly ready to be welded.

We are now ready to begin welding. Weld all the corners up and weld the 1-1/2" frame to the door front. Move around as you weld to avoid building up too much in the way of local stresses which could cause some warping. The short pieces of 1-1/2" square tubing will need to be drilled and welded into place before you begin assembly of the locking mechanism. These were, frankly, an afterthought. But they are necessary to keep the locking pins travelling in a straight line as the wheels turn. They may be seen in Photo 7. I used short pieces of the 1/4" wall tubing but a block of CRS or other steel will serve as well or maybe better.

To construct the locking mechanism itself begin by machining all of the required parts. Refer to the drawings for detailed information about the parts you will need to build the hand wheel, the hinges, the locking discs and their mounting hardware. The photographs in Series A show some of the details of their assembly. Here is where you will want to get innovative in building your safe. The arrangement of the rotating discs driving the locking plungers is one with many variations. A small safe with just one disc; a two disc arrangement; locking plungers on all four sides of the door; all of these are possibilities and there are many more. Figuring the length of thrust of the locking plungers relative to the location around the periphery of the operating disc is just a matter of laying it out on your table and calculating the result. A penetration of no more than the diameter of the locking plunger into the sides of the vault is sufficient. You may also choose to make bronze, lubricated bushings for the plungers to slide in, use hardened locking plungers or a combination of all of these variations. The results you end up with will be up to you and the amount of time and effort you are willing to put into it. Photo 8 shows the completed locking mechanism, including the combination lock, on this vault.

Building a Gun Safe

Building And Insulating The Box

After you have completed the fabrication of the front frame, the door and the locking mechanism, the fabrication of the box itself is pretty straightforward. But here is a good place for a reminder — this thing is getting pretty heavy! The assembly so far, if you have built it as described in these pages, is approaching 350 pounds! And that is just the front frame and the door! Here is where you will need either some sort of handling device or a whole bunch of willing teenaged boys. For my money, the load handling device is much easier to control than the teen aged boys would be. However you choose to deal with the weight please be careful!

Photo 6. The door and front frame assembly ready to weld. Move your welding around on the assembly, a little on one side then on the opposite, so as to minimize the effects of its tendency to warp from the welding heat.

That said, let's get on with the job at hand. If you choose to use the casters and/or the leveling screws listed in the bill of materials now is the time to add them. The casters will be useful during the construction process even if you choose to finish the job and remove them after the vault is placed in its assigned place. The leveling screws are not needed unless the box is going to be set up where the floor is unusually uneven. Photo 9 is a picture of the bottom front corner of the box with the casters and the leveling screws in place. A good procedure is to drill the bottom plate and the bottom piece of 1-1/2" angle iron to match the mounting holes on the caster plate, bolt the casters into place, and then weld the bottom plate and the angle to the front frame. I will not advise you here about how to best effect this assembly. But I will repeat — over and over if it will help —be careful while

Photo 7. A short piece of tubing welded in to help the locking pin move in a straight line. A block of steel will work here — maybe a little better.

Building Shop   115

Building a Gun Safe

Photo 8. A view of the box showing the arrangement of the locking mechanism as well as the beginning of the installation of the lining.

handling these heavy parts! I don't know about you but I don't have many valuables in my safe which are worth even one of my fingers!

Photo 10 shows the relationship of the corner angles to the top plate and the front frame. By assembling the components in this fashion you accomplish a couple of things. In the first place the parts, particularly the plates, do not have to be cut exactly to size. One rule many people forget is that when fitting up parts preparatory to welding a sloppy fitup is often better than a close one. It is easier to cut parts when exact sizes are not involved and it also will frequently make for a better welding job. The corners of the angle iron in this job were coped on a band saw and the plate was purchased sheared to size. Sheared sizes were specified to be about 1/4" smaller than the place being fitted to. Welding was easy, even though there was quite a bit of it, and the corners came out with a nice rounded profile.

Although I won't tell you how to sequence the assembly process I will make a few suggestions. Or tell you some things to avoid, as the case may be.

- If you can attach the casters and the bottom plate while the front frame and the door are still on the welding table that is a good thing. But be aware that if the casters are located much behind the front edge if you stand it up and then open the door it will over balance. (Would you like to know how I know that?)

Photo 9. The leveling screw and the top of the bolts attaching the caster. The leveling bolts are long enough to reach below the casters. If leveling bolts are required but casters are not used the bolts may be much shorter.

- It may make it easier to weld in the 1/4" bolts for installing the insulation and the lining if you don't weld all the sides on first. You might even consider welding the bolts to the plates before attaching them to the frame. I didn't do this but it would have made welding the bolts on easier.

- The casters on the vault shown here were removed after placing it in its final location. The box was then leveled up with the screws and the entire assembly was grouted into place. How you do it will depend upon where you put it when it is finished.

- Be sure to use either fiber lock fasteners or LocTite when assembling the locking mechanism. Being unable to open the vault because a nut worked loose and allowed the linkage to come apart would be what some people might call a "bummer".

Installing the insulation and the carpet lining is not difficult. Photo 11 shows the

Photo 10. Fitting up the frame for attaching the side plates. This makes for a good weld.

Building a Gun Safe

Photo 11. Beginning to look like a safe. Along with its extraordinarily handsome creator!

beginning stages of the process. I regret that I didn't take enough pictures of this stage of the construction but I was probably either getting in a hurry or my helper, in this case my son, was getting ready to go back to his home in Fort Worth, Texas, and we needed to move things along. But in the absence of photographic demonstrations I will try to describe what needs to take place:

- Begin by welding the 1/4-20 bolts to the walls of the box. Unless you just want to there is no need to measure and place them precisely. Create a pattern on each wall that will support the edges as you can see in Photo 8.

- Cut the bats of insulation to length and place it on the wall you are working on. Push it onto the bolts which will keep it in place until you are ready for the next step.

- Cut a piece of paneling or plywood, 1/4" is thick enough, to fit the wall you are working on. Then put the panel in place and tap it with a rubber mallet at each bolt. This will mark where you need the holes.

- Cut a piece of carpet the same size as the plywood

panel. Many carpet places will have scraps large enough to do this job so you should check with them before buying your carpet.

- Glue the carpet to the plywood and allow it to dry. A good grade of carpenter's glue in a caulking gun will do what you want. Put a bead of glue around the outside edge and put enough glue in the center of the panel just to keep things together until you get it bolted on.

- After the glue has had time to dry, place the panel, carpet side down, on a work surface large enough to accommodate the entire panel and, using an arch punch, punch 1/4" holes at each bolt location.

- Bolt the completed panel into place using large area washers.

After lining all of the sides, back, top and floor, you should make a panel to cover the inside of the door but it needs to be constructed in such a fashion as to not interfere with the workings of the lock mechanism. I used a 2 X 2 as may be seen in Photo 13. You will probably come up with a more effective and easier method for installing the lining than I used.

The lock is one which was available at a local locksmith's. There are many variations you might choose and sometimes people are sensitive about the details of their safes and the locks on them so I don't elaborate here. You will find that your local locksmith will be happy to help you choose a lock and will advise you as to its installation. But don't skimp on your selection. The work you have put into this vault up until now will require that you select an appropriate lock.

Furnishing the inside of the box has many possibilities. The box described here is large enough for you to install a filing cabinet or any other purchased storage shelf or unit. I have a two drawer filing cabinet in my safe and that works well. Gun racks or shelving of your design, either free standing or

Photo 12. Iinstalling the insulation.

Photo 13. The finished interior. Notice the 2 X 2 frame on the door. This keeps the insulation from interfering with the operation of the locking mechanism.

Building a Gun Safe

Photo 14. Finished and on the way to its new home.

I once heard Daylight Savings Time compared to making a blanket longer by cutting a piece off one end and sewing it back on to the other end. Makes just about as much sense.

fastened to the walls of the unit, have almost limitless configurations. If you are careful in your planning you might consider welding fasteners to the walls of the box before you install the lining which will help you to arrange the interior of the box. I fitted the inside of my original safe with special shelves and fasteners and then almost immediately traded and swapped the guns I built the shelves for and had to redesign the inside of the box. So keep things flexible and if you trade for a 40mm mortar you will have a place to keep it without having to build a whole new safe.

And that pretty well does it. You will vary your approach to building a safe or a vault according to your shop capabilities, your material availability, your time and the extent of the fortune you need to protect. If any of you are old enough to remember the old Disney comic books you will no doubt remember Uncle Scrooge's money bin. I don't have that much in terms of material wealth to protect but I don't want to risk losing the items I do consider valuable. I keep the Bulgin family *Bible* in my safe. It was purchased by my great-grandfather in 1861 and I like knowing that it is protected against either theft or fire. And if I ever make a million dollars this safe will be a good place to keep it as well. I would probably spend most of the million dollars on adding to my shop and the tools in it so having a safe isn't that necessary for me. Except for a place to keep the family *Bible*.

Series A Photographs

Photo A1. Three stages of fabricating the locking discs. On the left are the 3 components of the wheel before welding. In the center is the wheel welded up and on the right is the finished wheel.

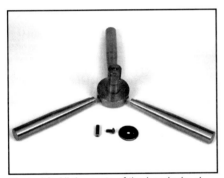

Photo A2. All the parts of the hand wheel before welding.

Photo A3. Component parts of the leveling feet. These are optional and not really necessary. I knew a little about where this vault was eventually going to be placed so I added them.

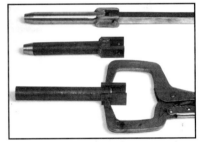

Photo A4. The steps of fabricating the locking plungers. The tabs should be welded on before drilling the holes for the linking bolts.

Photo A5. Required parts for fabricating the hinges. This hinge is easily fabricated and will handle about all the loading you care to put on it.

Photo A6. Setup for welding the hinges in place. These hinges may be purchased but fabricating them is so easy I didn't want to wait for UPS to bring them.

Photo A7. Right, machining the welded locking disc. Hot rolled 1/4" plate welds beautifully — machines terribly!

Photo A9. Below, setup for drilling the equally spaced holes around the hub of the hand wheel. The holes make it handier to weld the assembly together as seen in the next photograph.

Photo A8. Machining the taper on the hand wheel spokes. Set the compound at about 3-1/2 degrees and you will have an eye pleasing taper.

Photo A10. Hand wheel components assembled and ready to weld.

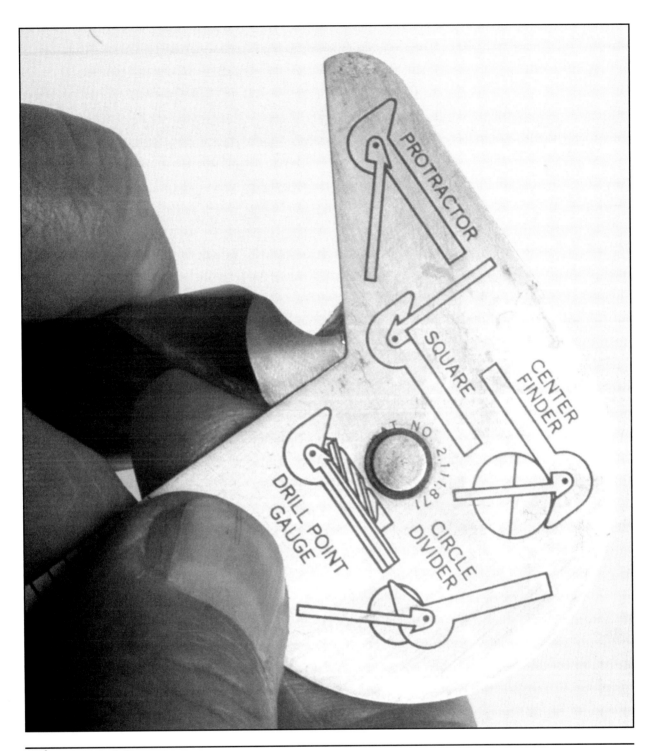

Chapter 10

The Care and Feeding of the Twist Drill

Twist Drills

The thing about the common twist drill is that there is nothing common about it at all. It is a complex, effective, cunningly designed cutting tool which is found by the thousands in every corner of the world. It is largely taken for granted yet if it were not for this ubiquitous little cutting tool, the machining of almost any part we make would be vastly more complicated. I am going to talk in this chapter about the twist drill and I am going to give you a lot of information, most of which you will already have and a great deal of which you may not even want.

I have heard many home craftsmen say about their shops at home that they have this or that piece of machinery and tools that will do such and so and that they have a "full set of drill bits." I never dispute them. You never know really who you are talking to and they could actually have a "full set of drill bits." But I am here to tell you that we are talking about a whole bunch of drill bits! I have done a little figuring on the subject and here are the results. Keep in mind that I am talking only about ordinary drills ½" in diameter and smaller.

- Fractional drills from 1/16" to ½" by 64ths — 29
- Number or wire size drills from #1 through #60 — 60
- Letter drills from A to Z — 26
- Metric drills — 1mm through 13mm by .5mm increments — 25

We have so far 140 drill bits, all different except for the E and the ¼" sizes. Now before you start writing me letters let me say that I do realize that there are also common number sizes of 61 through at least 92 and that you can get fractional drills smaller than 1/16". I also know that a 13mm drill is slightly over ½" and that metric drills may be obtained in .1mm increments so coming up with a truly accurate count of drill sizes would be daunting. This is maybe a little conservative but it is a good starting place.

Now let us go to drill lengths. I have drills in my shop in screw machine length, in jobber length and in extra length. For the sake of our accounting here we will limit drill lengths to those three classifications although there are others. We will ignore aircraft drills, taper length drills and other available but not widely used configurations and use just these three standard lengths. Using our figure of 140 drill sizes and multiplying that by the three available lengths we are now up to 420 different drill bits.

But we aren't finished yet. All of these drills are also available in left hand spiral which brings us to 840. And you can buy any

I agree that we are better off without some of the chemical compounds like trichloroethane, carbon tetrachoride, and white lead. But sometimes I sure do miss what they did for us.

one of these drills in either bright, black oxide or TiN (Titanium nitride) finish so we have to multiply again by three bringing the total now up to 2,520 different configurations of drill bits. Oops! Forgot about points. We have available to us as point choices either 118 degree, 135 degree, split point or flat points. That takes us up to 10,080 choices and remember we are still staying under 13mm which is about .012" over ½" in diameter.

And, oh yes. There is one last thing worth mentioning. Most of these tools are available in HSS, cobalt or solid carbide and there are other options when the various coatings are considered. So all told we are looking at over 30,000 different, legitimate configurations of cutting tools. So you will understand why I roll my eyes when somebody tells me that they have a "full set of drill bits" in their shop.

The purpose of this chapter with its accompanying photographs is to discuss drill bits. It is a fascinating subject if you are as shallow a person as I am. I find it far more interesting than discussing world affairs about which I can do nothing. And politics isn't even in the race as far as I am concerned. Who could possibly find anything remotely as interesting in a panel discussion about the current state of affairs in some far off land as an in depth consideration of a twist drill? I ask you?

Photographs number 1 and 2 are pictures I am including to emphasize the fact that we take the twist drill in all of its various configurations for granted. This is a drill which was made by my grandfather. He died in 1936 so I have no idea as to the particular purpose or job for which it was made. It is 1.730" across the flutes. Notice that there are no relieved margins along the flutes of the tool. It was apparently made by forging a tapered flat and then twisting it while hot. The number 2 Morse tang was welded on, probably after the flutes were formed. I cannot tell from my examination what method was used to weld the tang on but I know from having talked to people who knew him that my grandfather was a wizard at forge welding. He served a five year apprenticeship in Chicago as a blacksmith in the 1890's and then worked as a blacksmith until his death in 1936 so he had time during those years to get to be good at what he did and all the evidence I have proves that out. You can tell from the photos that the tool had all of the characteristics we look for today in a good drill bit. The cutting lips are of equal length. The chisel edge or cutting point is clearly defined and at the proper angle to the cutting lips. The clearance is uniform on both flutes

Photo 1. Top, a hand forged drill bit from the 1920s. 1.73" in diameter with a #2 Morse tapered shank.

Photo 2. Above, the point of the forged drill. Notice the clean and accurate lines of the cutting flutes.

Twist Drills

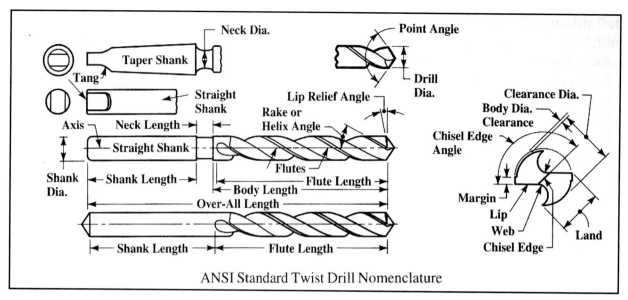

Photo 3. A drawing scanned from *Machinery's Handbook* showing the nomenclature of the twist drill.

and is at approximately 12 degrees. This was likely a good tool for its day. I have never tried using it because my grandfather's shop burned down in 1935, an event which likely contributed to his death a year later at the age of 61, and this tool, along with many others, was recovered from the ashes, its temper and usefulness gone forever. But to look at this cutting tool and to know that today I can pick up the telephone and call an 800 number and order any size or quantity of drill bits, of any configuration we have listed here, and expect it to be delivered to my shop tomorrow, really makes me realize how much for granted we take these tools. Everything about the good ol' days wasn't all that good!

But let's get back to the subject. I have scanned from *Machinery's Handbook, 26th Edition*, a drawing of a twist drill, Photo 3, which provides a pretty complete reference for our discussion. As a matter of fact, it is a little too complete which can be confusing to the beginning machinist. But there are a few critical things to consider when selecting, using or sharpening a twist drill.

- **Diameter** — Obviously you will be concerned with diameter and it pretty well explains itself. You will on occasion need to drill a hole smaller than the finished size. You will ream or bore to size depending upon the level of precision required so you start out by drilling undersize an appropriate amount. Most of the time, however, your choice of drill size will be the size of hole you need.

- **Shank** — Both shank size and configuration are important and the type chosen in most cases will be either Morse or straight. A reduced shank is sometimes necessary in order to drive the tool with a chuck whose capacity is smaller than the drill diameter. But a word of caution is in order here. Just because your chuck will accept a tool does not mean your spindle can drive it. It is possible to machine the No. 5 Morse shank of a 2" diameter drill bit down so that it will fit your ½" Jacobs chuck but I don't think you will want to do it. A good option to have on some Silver & Deming drills is three flats machined at 120 degree points around the shank to help prevent the chuck from slipping. An example of this may be seen in Photo 4.

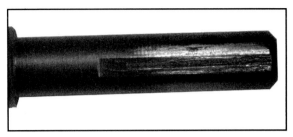

Photo 4. One of three flats machined on the shank of a twist drill. This is a good way to avoid having the drill turn in the chuck which results in a damaged shank.

- **Point angle** — The included angle of the point of most of the drills available at the local hardware is 118 degrees. This is a good general purpose angle and your drill point gage will help you to maintain it when sharpening the tool. But don't get hung up on it. If both flutes are the same length and the same angle, then the actual angle is not as critical. There will be times, depending on your material, when other angles will work better.

- **Lip length** — This is a feature not clearly shown in Photo 3 but which is important for this reason. The lengths of the lips should be as nearly exactly equal to each other as possible. Otherwise you will likely drill an oversize hole.

- **Lip relief angle** — Another feature whose presence is more important than its value. If there is no clearance the back edge of the cutting lip will contact the work either before or at the same time as the cutting edge and the result will be a lot of heat, a damaged tool and probably some bad language on the part of the operator. I have seen more examples of inadequate clearance angles than you would think considering its importance.

- **Flute length** — Of course your drill will have to have flutes long enough to drill to the depth you require. A rule I have tried to follow, I have no idea of its origin, is to try to avoid designs where a hole must be drilled deeper than three times the diameter of the tool. You cannot always follow this rule but when you need to drill deeper, withdraw the tool often to get rid of chips which will build up in a deep hole.

Other features of drill bits are important in their place but if you make your selections by considering the ones listed here you will usually be able to do the job. If you look at one of the major tooling catalogs, MSC for example, you will find that the available choices are seemingly endless. But keep to the basics. Consider the characteristics of the hole you are going to drill, the type and condition of your material and the machine you are going to use for the job. If you select a tool which fits all of these conditions you won't have any trouble. You believe that, don't you?

What Drills Do You Need?

What drills do you need to keep in your shop? You, obviously, will be the only one with enough information to answer that question but there are a few guidelines. If the only hole drilling you are called on to do is to fill the requirements of home maintenance then an index of jobber length, HSS, fractional drills will likely do it. Maybe even only up to 3/8" in diameter. If you also do the maintenance on your automobiles you should add an index of numbered drills from #1 through #60. But if you are reading this there is a strong likelihood that you will need, or want, a more complete selection from which to choose.

When I was 30-years-old I was a foreman on a welding crew. There was a 45-year-old weldor on the crew who used a 1.5X cheater lens in his hood. We thought that was pretty funny!

I could give you a list of what I keep in my shop but it would take a long time and you would think I was bragging so I will forego that. But here is what you must consider. If you have customers who depend upon you to provide quick service you not only need, for example, a #7 drill which is the tap drill for ¼"-20 threads, you also need to keep one or more spares of that size. The same thing is true of other tools which get frequent use in your shop. If you are only working on your hobby projects, and don't think I am minimizing the importance or the priority of those projects, you might better be able to wait until you can order a new bit to replace the one you just broke. It depends on your own priorities and I can't advise you there.

I can, and do, recommend that you get a set of screw machine drills. Maybe a set of each, fractional, number and letter sizes. I use those drills in my milling machine almost exclusively because they don't require as much adjusting up and down of the knee of the machine.

Sharpening the Twist Drill

I have seen many attempts, and you have too, at describing the methods and procedures for the offhand sharpening of a twist drill. Some of them are good, all of them are well intended, but most of them, and I include what you are about to read here, fall short of the mark. There is no substitute for a qualified teacher first demonstrating the process and then watching and correcting you while you try it. In my case it was a Machinery Repairman Chief Petty Officer at the U. S. Navy school for machinery repairmen in San Diego, Calif. A long time ago. I have had the opportunity to teach many others in the years since then, some who got to be good at it and some who should have stayed in the clerical trades. But I am going to try to show you here with some pictures and some words how to do it. If this isn't helpful to you then I will accept the blame. Refer to the series of photographs A through J.

First you must acquire, or make, a good drill gage. Good examples of a purchased gage and a homemade gage are shown in Photo A. The homemade gage is merely a piece of 16 ga. brass sheet with a 59 degree angle cut into it. It will serve just as well as the purchased gage on the left except for the fact there are no graduations on it for comparing cutting lip lengths. A machinist's scale will serve that purpose. I have added a feature to the home made gage which is not included on the purchased gage. Across the bottom there is cut an 11 degree angle which serves as a clearance gage. This too, you will soon learn to judge by eye without using the gage.

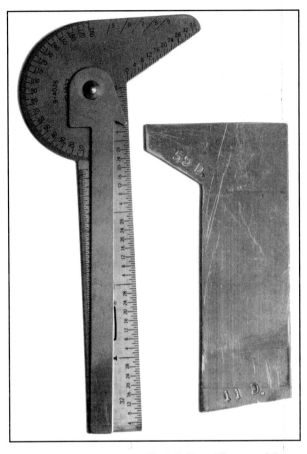

Photo A. Two drill gages. A "store bought" on the left and a homemade on the right.

Photo B. The proper position for approaching the wheel to begin the sharpening process.

You will also need a good bench grinder with a firm and square tool rest and a true running, aluminum oxide grinding wheel of about 46 grain size which has been dressed to a clean, square surface. This is a good selection for a wheel for use in the offhand grinding of HSS cutting tools. I won't tell you it is the only one but it will do the job. There are some pretty good synthetic grit wheels now but I don't have a great deal of experience with them. If you are grinding larger drills you will, obviously, need a wheel with a wide enough face to accommodate the entire length of the cutting lip.

Begin by bringing the cutting lip of the drill to the wheel as shown in Photo B. I usually keep two fingers under the tool but one finger will work. On smaller tools one finger works better.

Twist Drills

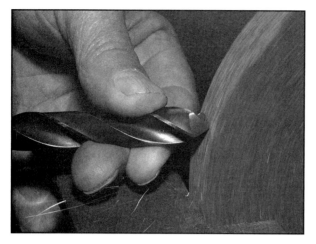

Photo C. Move the end of the drill in the right hand down *without* rotating the tool.

Photo D. Continue moving straight downward, holding pressure against the wheel.

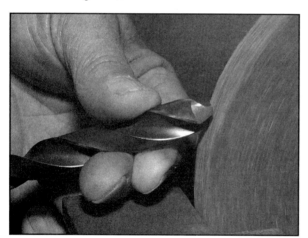

Photo E. Coming off the grinding wheel.

Photo F. Checking the lip angle and length with the drill gage.

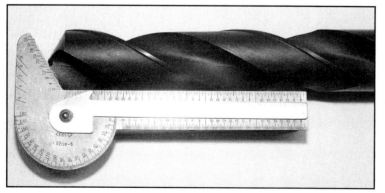

Experience will soon teach you the proper angle but until you gain the experience, compare the tool to the gage frequently.

Photos C, D and E show the sequence of movements to make while grinding the tool. IMPORTANT! Do not spin or rotate the drill as you grind across the face of the point! This is the hardest thing to get across to the novice as there seems to be a natural tendency to do this. After making contact with the grinding wheel, move the back end of the drill straight down with your right hand while keeping the tool against the grinding wheel with your left hand.

If the drill you are grinding has a lot of material to be removed, say it has a chunk broken out of the point, you may take two or three passes on one flute before turning it but I recommend that you alternate between the faces. That way you will keep the lip lengths more even. Unless you are grinding really small drills, do not use any sort of chuck or pin vise to hold it because you run a greater risk of burning the tool. Your bare fingers will tell you when it is time to cool the drill between passes against the grinding wheel.

Check fairly frequently against the gage as seen in Photos F and G. Also check visually between passes to see that you are maintaining the

Photo G. Checking the clearance angle.

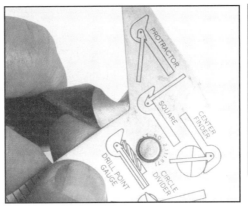

Photo H. Checking the angle of the chisel edge or cutting point of the drill.

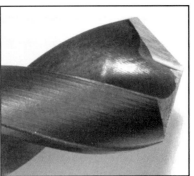

Photo I. A drill modified for drilling brass. This will prevent the drill from digging in.

proper clearance angle behind the cutting lip. The final proof of whether or not you have done the job correctly is in the pudding and in this case the pudding is a hole you drill. A properly ground drill bit will bring out two identical shavings and the finished hole will be almost exactly the size of the drill you are using.

While we are at the grinder let me mention a couple of other processes you will find useful. Photo I shows a cutting lip modified to drill brass. If the rake angle, or cutting angle of the tool is left as is, there will be a tendency for the bit to grab the work piece and pull itself into the work. The lip is modified by changing the cutting angle making it less acute. Photo J is an example of web thinning. As drills are reground and "used up" the tapered web becomes thicker as it gets shorter. This can be corrected by grinding away some of the web on both sides.

Photo J. Another modification. The web has been thinned by grinding.

Photo K. A small drill grinding machine. Limited to ¾" and smaller drills but a useful machine in a small shop.

A discussion of drill bit sharpening would not be complete without including a mention of the drill grinding machine. For grinding small drills in quantity it is hard to beat a Black Diamond. It uses a system of individual collets for each drill size and gives a high degree of accuracy and concentricity. Rush is another good machine, particularly in the larger sizes of standard twist drills. I use a Darex machine in my shop and it is probably one of the best values for the small shop. It does a good job on drills, both standard and split point, up to ¾" in diameter. It has Borazon/CBN grinding

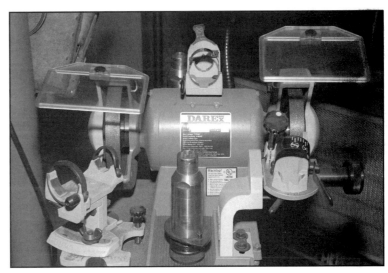

Twist Drills

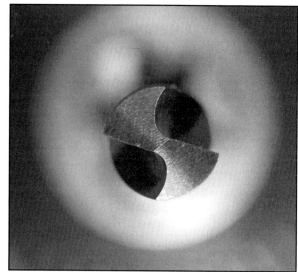

Photo 5. The point of a 118 degree ½" diameter drill. Right hand twist.

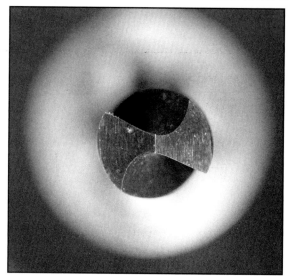

Photo 6. The point of a 118 degree ½" diameter drill. Left hand twist.

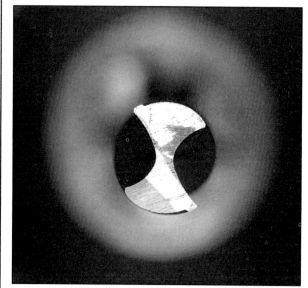

Photo 7. The point of a ½" diameter drill with a flat point.

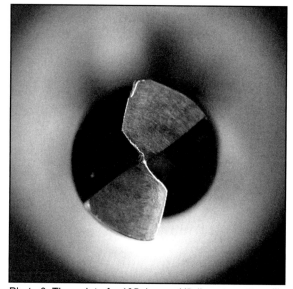

Photo 8. The point of a 135 degree ½" diameter with split point.

Photos 5-7: The business end of four different drill grinds.

wheels and I cannot tell you how long the wheels last because I have never worn one out. I could get along without a drill grinder at all but I do love machinery auctions and some of the tools available at them. Photo K is a picture of my drill grinder.

I hope this attempt at describing a fairly simple, yet often misunderstood, process will help somebody acquire a new skill. Or maybe improve an old one. But if you still have questions or if you feel that you are still doing something wrong then feel free to come by my shop and I will show you first hand. That is the best that I can offer.

Twist Drills

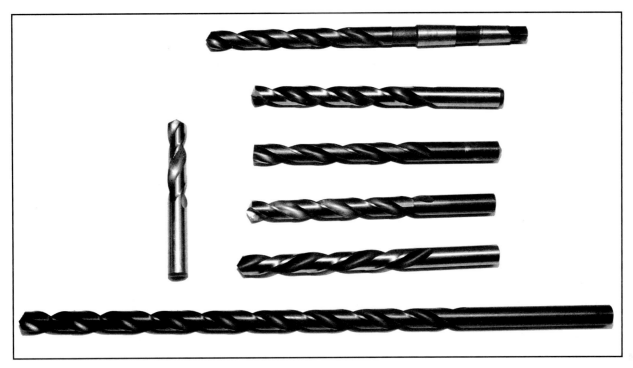

Photo 9. A "family portrait" of 27/64" drills of different lengths and styles.

Some Variations

I am including a couple of additional photographs here just to demonstrate to a little greater degree the amazing versatility of the drill bit. The sort of spooky looking photos numbered 5, 6, 7 and 8 are head on pictures of the business end of four different drill grinds. Photo 5 is a standard, 118 degree ½" drill bit. Photo 6 is of a left hand spiral drill. Photo 7 is a drill ground to a flat point while Photo 8 is a 135 degree split point drill — which I didn't realize had a chip out of one flute until after I took the photo! Photo number 9 is a "family" of drill bits. They are all 27/64" in diameter — tap drill for ½" -13 threads --- in seven different configurations. Horizontally from bottom to top they are: extra length, left hand spiral, right hand spiral, flat point, 135 degree split point and #2 Morse taper shank. The short drill to the left is a screw machine length drill. I use 'em all! In Photo 10 you see two examples of the Silver and Deming configuration drill. These drills are particularly useful in the small vertical milling machine. And Photo 11 is a 33/64" diameter drill which has been modified to drill a 3/16" hole and then counter bore a 33/64" diameter all in the same operation. There are many such applications where the ordinary twist drill can become an extraordinary cutting tool.

Photo 10. Belw, two examples of Silver and Deming drills.

Photo 11. Bottom, a 33/64" drill modified to drill and counter bore a hole in one operation.

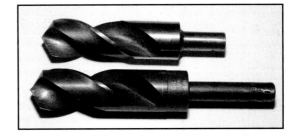

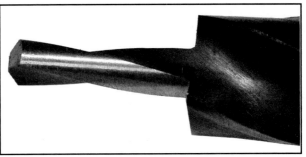

Building Shop

Twist Drills

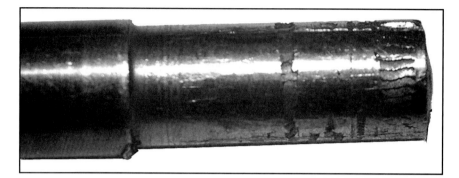

Photo 12. An example of shank damage done by being allowed to turn in the chuck.

A discussion such as this one would not be complete without a little preaching so I will preach a little. It irritates me to see dull, chipped or broken drills, cutting tools of all sorts, really, but it seems drills are usually the innocent victims, in an otherwise well run machine shop. Or a garage or anywhere drills are used. I also hate to see what I call snaggle-toothed drill indexes. An index for fractional drills has 29 holes in it. There should be 29 drills in those holes. Here is the rule — When you are finished using a drill bit it is in one of three conditions:

- **Condition one:** It is still sharp and serviceable with an undamaged shank and should go back into the index right then. It is OK to wait until the job is finished but when dark comes all of your drill bits should be at home.

- **Condition two:** It is dull or chipped and/or the shank has turned in the drill chuck scoring the shank. In this condition, see Photo 12 for an example, it should be resharpened and/or the shank dressed and then put back into the index. (The split point drill you saw in Photo 8 has been attended to.)

- **Condition three:** It is damaged beyond economical repair and should be disposed of. Get rid of it. There is no market for unusable drills. If you keep them around they could actually get you into trouble by making you think you have something you do not have.

There you have it. A lot of words about an unassuming little cutting tool and I have only scratched the surface. My father had a ½" Craftsman drill press in his shop when I was growing up and it was the first power tool he let me use. I was lucky that I learned to clamp my work to the table and tighten the chuck securely and select the proper drill and still keep all of my fingers and both of my eyes. I owe him for that, by the way. I drilled my first hole over 60 years ago and I probably burned up my first twist drill within a week of drilling that first hole. I still have a lot more of both to do.

They say that we learn from our mistakes. If that is true I must know about everything there is to know by now.

Chapter 11
Some Things You Need in Your Shop — and You May Not Even Know It

Building Shop

Some Things You Need in Your Shop

It is presumptuous of me to be telling you what you need in your shop when you are the only one who really knows what your shop is or what you do in it. But the items described here are in a peculiar category. Take the first one, the burning table, as an example and you will see what I mean by peculiar. You can get along very well without one of these. You can use saw horses or an upended 55 gallon drum and do a lot of burning without the benefit of having a dedicated table. But if you will take the time to build one you will be amazed at how useful it is, not only for its primary purpose but for many other jobs as well. Most of the items shown here are similar in that respect. If I didn't have my chip hook I would break all of my fingernails twice a year removing the top of the box where the water valve to my outside faucet is located. The diamond wheel dresser is pretty limited in what I can do with it but I wouldn't want to be without it. So I guess I am saying that these items are things I need in my shop and who am I to make suggestions as to what you need or don't need in your shop. And if I set my mind to it, or if you do, I'll bet we could come up with a couple of dozen things in the same category. Look at these suggestions and you will see what I mean.

Burning Table

The first item is a burning table. The old warhorse shown in Photo 1 was built by my brother in 1963. He had used one like it in the boiler shop where he worked in Seattle so he came back home and built this one. Its simplicity is its most useful asset. It is easy to build and except for the hot rolled strap from which the top is made the whole thing is made from things you can get at nearly any automotive junkyard. You can even use discarded leaf springs for the top if you want to go to the trouble. It can be built in half a day easy and if you do any freehand burning you will never regret spending the time it takes to build it. It also serves as an ideal welding positioner for small jobs. All of the locking discs were burned out for the safe locking mechanism we talked about in Chapter 9.

In one of the construction articles

Cover Photo ---- Completed burning table.

Photo 1. The top of my 33-year-old burning table. This guy has seen lots of use and it shows it.

in *Home Shop Machinist* magazine the writer described the process for building his project and he advised strongly against using a cutting torch to produce some of the parts for the job. That told me something about the writer. He was not, and likely had never worked with anyone who was, experienced in the use of an oxy-acetylene cutting torch. He was placing the blame for the inability to master a process on the process itself. In the years I have spent working in pipe shops and in fabrication shops, I have known craftsmen whose abilities with a cutting torch were just hard to believe. I was never as good as the best I have known but I did learn pretty well some of the basics of using the oxy-acetylene torch and the most basic of the basics is to be in a comfortable position when you are burning with the torch. And this table accomplishes that for everything you need to burn out unless you are building large earth moving machines. Or maybe sea going vessels!

Material List, Burning Table

1. 20' or thereabouts of 3/8" X 1-1/2" HRS strap. 1/4" thick might work if you only plan to use the table infrequently but I recommend at least 3/8". I have welded some pretty heavy components on my table which justifies the heavier material. Old leaf springs from the junk yard can work if you want to go to the trouble of taking them apart and if you can ignore the curves in the material.

2. A front wheel spindle from a junk yard. Be sure to get the hub and the bearings with it. Get one from a half-ton pickup truck or a mid size car if you can but even the small ones will do OK.

3. A piece of drive shaft from the same junk yard. 2-1/2" to 3" in diameter and long enough to make the column.

4. A 14" or 15" wheel. One of the "polio" wheels used these days as a temporary spare will work well enough.

5. A couple of scraps of HRS plate.

If you decide to build one for yourself the only specification which is really important is the height. Experiment some as to what the best height is for you. I am 5'11" tall and a table 40" high is perfect for me but you may have other preferences. And remember as usual that the materials list here is provided as a general guide only. Make substitutions according to your needs, your abilities and equipment availability and most especially, your pocketbook.

Begin, appropriately, with the cutting torch. Although you will have to make do with the saw horses for a burning table until this job is finished. Burn the steering arms and the shock mounts off of the spindle until you are left with just the spindle and a flange.

Photo 2. Automotive spindle. Trimmed and ready to weld.

Some Things You Need in Your Shop

Photo 3. Welding up the new table top.

Photo 4. Spindle welded and ready for assembly.

Photo 5. Spindle and hub assembled.

Then weld the flange itself to a piece of plate about 4" square. See Photo 2.

Now make the table top itself. You will have to roll a piece of the 3/8" X 1-1/2" strap into a circle about 24" in diameter. I formed this one on a Hossfeld No. 2 bender. Weld the ends so that you have a 24" diameter ring. Lay out on the welding table a series of parallel lines as seen in Photo 3 and cut the cross pieces to fit into the ring with 1-1/2" spaces between them. When you are finished welding up the table top weld the completed spindle assembly to the center of the top. Refer to Photos 4 and 5.

Cut the drive shaft to a length which will give you the height you want and weld one end of it to the wheel and the other end to the spindle. Yes. You are going to weld the spindle assembly so that you will never be able to service the bearings but that won't be a problem because they will never need servicing. You will have packed the bearings with new grease and made sure that the cotter pin is in place and the nut tightened to the proper adjustment. And if they ever do require service, you will have used the table so much by that time that it will be ready to be discarded and you will be building a new one anyway. Only the next time you will have the old one to use in the process. After you have finished with building this accessory you will wonder how you ever got along as long as you have without having one.

Circle Burning Attachment

The second item here is one which will complement, or will be complemented

Photo 6. Circle burning attachment mounted on cutting torch.

by, the burning table. A circle burning attachment for your cutting torch. I own a Victor cutting torch which I purchased in 1964 for $93. This same torch is still available today but now sells for well over $200 plus the cost of the tips. I made the circle burning attachment shown in Photo 6 soon after I bought the torch and it has served me well for all of the intervening years. Photo 7 is of the component parts of the attachment and Photo 8 is of the circle burning attachment and the burning table in use. Note that the attachment does not require any mechanical clamping onto the torch body. The small circle that fits over the tip of the torch is attached at a very slight angle so that it holds itself in place when clipped to the body of the torch.

Photo 7. Circle burning parts.

I cannot provide a material list for this job for obvious reasons. Your torch will likely be different from mine (You should be so lucky as to have a cutting torch as good as mine!) and there are other variables. But use this as a guide and make one of these attachments for whatever cutting torch you own and you won't be sorry. You can purchase generic burning aids for your torch but usually there will be sacrifices in quality caused by the attempt to make them universally adaptable to every type of torch. You will be happier with one you have designed and made to fit your own application.

Chip Hook

The next item on the list of things you cannot do without is a homely little device which has usefulness completely out of proportion to the effort required to make it. A chip hook. I call it a chip hook because that is what I made it for but it has hooked about everything. I use it to retrieve items from the

Photo 8. Burning a circle using the attachment and the burning table.

Building Shop 139

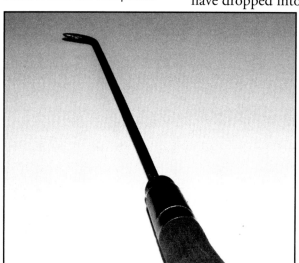

Photo 9. Back end of chip hook.

slack tub. I use it to drag parts out to the front of the heat treating furnace so that I can reach them with the tongs. I retrieve things I have dropped into the chip tray, providing I can ever find anything I have dropped into that bottomless pit. I probably use this hook more than I do a 10" Crescent wrench and that is a lot!

Making it is simple. All you need is about 2' of 1/4" round HRS, a file handle and a couple of washers. Drill all the way through the file handle and then drill a counterbore in the back end of the handle. Weld a washer onto the end of the 1/4" round and then push it through the handle until the washer is flush with the back of the handle. Then weld another washer to the ferrule of the handle and to the rod where it comes through. Sharpen the end of the rod and bend it to ninety degrees about a 1-1/4" from the end and there you have it. The photographs show it better than I can tell it. Scratching your back in that hardest to reach of all places is just one more perk from this little tool. Photo 9 shows the way the washer is fitted into the back end of the handle and Photo 10 is of the completed tool. A fairly sharp point is a good thing to have on this tool so that you can get under that ball of chips which persist in forming on the end of the boring bar.

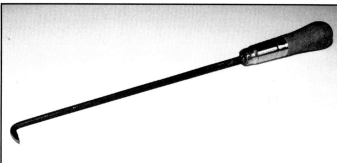

And I can hear the objections from here! But if you have your tool and your feeds and speeds all perfectly set up for the job the chips won't ball up on your tools! Sounds good to me but if anyone reading this can make the claim that they never have any problems

Photo 10. Chip hook.

with chips, particularly on the lathe, then you are invited to come to my shop and we will talk about it. The only way to avoid completely the problems associated with making chips is to not make any. And wouldn't life be miserable if that were the case?

Speed Wrench

The next item is a speed wrench for a three-jaw chuck. My scroll chuck takes 54 full turns of the chuck wrench to open it up from 1" to 6". And the jobs which come into my shop are never in order. That is, I will get a job to make an extension for a 3/4" diameter ship auger and the next job will be to machine out the bore of a 5-1/2"

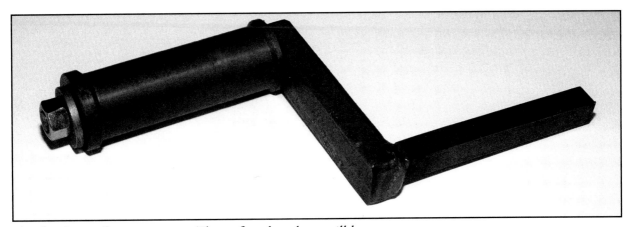

Photo 11. Speed wrench.

chuck adapter for a customer. Then after that there will be someone who wants threads cut onto a piece of 5/8" diameter drill rod. Next will come a 6" diameter blade hub from a riding lawnmower. And it is open and then shut and then open back up again. You won't change the order in which your jobs come so you must change the way you adjust your chuck jaws and here is how to do that.

To make this speed wrench for your three-jaw chucks all you need is a piece of key steel, or hex stock depending upon your chuck, the same size as the socket in the chuck where your chuck wrench fits and something to use for a handle. It is best if you make the handle so that it turns freely but that is not a requirement. Just a piece of round stock welded on will work nearly as well.

My every day chuck has an opening of just over 7/16" square so a piece of 7/16" key steel works perfectly. Photo 11 is of the speed wrench made to fit the chuck on my 17" Kingston lathe. I also have a similar wrench made to fit the chuck on my Hardinge HLV-H lathe. And if I had a third machine, or different chucks to fit these machines, I would have speed wrenches made for them as well. It is a tool you will use almost every time you use your scroll chucks. A four jaw independent chuck is a different matter but you might benefit from having a similar wrench to fit it as well. The speed wrench is not intended to provide enough torque to tighten the part in the chuck but should serve only to bring the jaws into position for tightening. Use the wrench which came with the chuck for the final tightening. And if I may be allowed to preach for just a moment — don't put a length of pipe on the handle of a chuck wrench to tighten it!

I feel a degree of sympathy for the person who has learned all there is to know about his or her profession. They probably don't even realize how much they still have to learn.

Length Gage for Your Cutoff Saw

Most of you who regularly use a cutoff saw in your shop have figured this one out already and will have made and installed one of these attachments for your saw. And many saws will have the attachment when they come from the machinery distributor. But it is so useful

Some Things You Need in Your Shop

Photo 12. Bottom of saw table showing mounting of stop.

Photo 13. Sawing shorter material.

an item that it will bear this repetition. Whether your work entails cutting spindles or pickets for hand railing or sawing slices from a piece of 6" diameter 6061 alloy aluminum you will be able to use this device to your advantage.

There are as many, probably more, variations of cutoff saw in our line of work as there are opportunities to buy cheap medications on the Internet so I won't be specific as to just how you should go about making this. But there are a couple of things to consider and the first thing is how you will go about attaching the stop to your machine. To be the most useful, an attachment should be designed so that it can be removed completely if it could get in the way of other operations. I machined the two collars seen in Photo 12 and welded them to the underside of the tip off table on my saw. When the gage needs to be removed all that is required is to loosen the setscrew and slip it out leaving the collars out of the way under the tip off table.

Photos 13 and 14 are of the two different ways that the gage can be used. If the pieces to be cut are shorter than the length of the tip off table then turn the gage so that it is in the position shown in Photo 13. For other parts it can be used as you see in Photo 14. The set collar is in place so that you can set your material for the cut and then move the stop away from the end of the work. This will insure that the blade will not be pinched when the cut is complete. To use the stop you must first position your material so that the cut will be to the appropriate length. Clamp the vise securely. Position the stop and lock the set collar to the shaft and then move the stop away from the end of the work piece or pieces. After the cut is made you can then slide the stop back into position against the set collar, move the material up into place and clamp the vise, and then move the stop out of the way again.

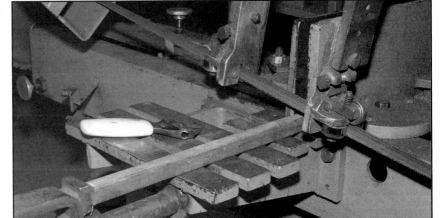

Photo 14. Sawing longer material using stop.

Photo 15 is of a similar device made to fit on my homemade abrasive cutoff saw. The principles are the same. This one is threaded into the saw table and it is moved to avoid pinching the blade by turning it CCW after positioning the work. Make the cut, turn the stop back into position, reposition the material and repeat the process.

Photo 15. Cutoff stop on abrasive saw.

When sawing several pieces of material to be machined in subsequent operations, particularly CNC operations, both accuracy and repeatability are important. If you are careful in the design, installation and use of a gage such as the one shown here, and if your saw and its blade are in good condition, you can routinely cut material with an overall length tolerance of less than .010".

A Diamond Wheel Dresser for Your Pedestal Grinder

I would venture to say that no tool in our shops, save perhaps the Crescent wrench, is more taken for granted and possibly, more abused, than the bench grinder. Or pedestal grinder, or whatever you call it where you are. It is the exception rather than the rule to go into a shop and find a grinder with the wheels true and clean and with the work rests square and adjusted with the proper clearances and relation to the grinding wheel. If yours are then please accept my compliments. One of the reasons for their neglect or their poor condition is that dressing the wheels on a bench grinder is a chore. When we need to use our grinders we need them right then. And when we are finished we need to get on with the job that necessitated their use. I worked in a large shop at one time where it was an apprentices' job, each apprentice served his time at this, to keep all the machine lube reservoirs filled and all of the grinding wheels dressed. Unless there is some sort of dedicated program of this nature in place the wheels most of the time just don't get dressed. Enter the diamond dresser!

Photo 16. The diamond wheel dresser for pedestal grinders. Note the knurled screw for use in incrementally adjusting the diamond.

Photo 17. Another view of the dresser. The tool rest must be modified so that there is straight and smooth surface upon which to move the dresser from side to side.

This device takes a little time to make, and you must also modify your work rests on the grinders in order to use it, but once you take the time to make these changes to your grinders the chore of wheel dressing is no longer a chore. It will become so easy you will never again neglect your grinding wheels. The pictures should be self-explanatory. Photos 16, 17 and 18 are three different views of the device I have made for use in my shop while Photos 19 and 20 show details of its construction. If you have more than one grinder in your shop, and you should, I have five I use frequently and three more just in case, then you will want to design your dresser so that it may be used on most of them with little modification. Photos 21 and 22 are of the tool rest I made for my large grinder and the dressing tool in use on a smaller grinder. The trick is to make this thing really easy to use. Your grinders will appreciate it.

Some Things You Need in Your Shop

Photo 18. Left, the black knob on top locks the diamond sleeve in place after placing it roughly in position. Then diamond is then moved out by the screw in the back.

Photo 19. Below, diamond dresser assembled.

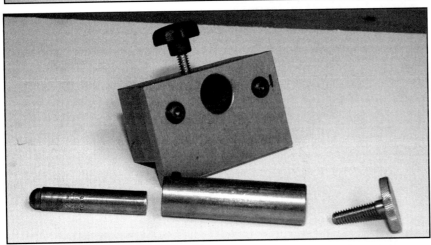

Photo 20. Component parts of the dresser.

Photo 21. Modified tool/work rest required for using the diamond.

Building Shop 145

The Tool Post Grinder

The last item on this list of necessities is not the requirement of the tool itself but rather a little different approach to the method of mounting the tool and how it is used in the shop. A tool post grinder gets a lot of use in my shop. And I hope you are not one of those people who will say that you should never make a practice of using tool post grinders on lathes because of the swarf and the grit which will get onto the ways of the machine and before you know it you will have ruined your lathe. Take the proper precautions to keep the swarf away from the areas of the machine where grinding dust could do some damage and clean up the residue after using it and you can use the grinder on your lathe for as long and as frequently as you need to.

But I am getting ahead of myself. What I am talking about here is the place that you keep the tool post grinder when it is not on the lathe. I have a Dumore tool post grinder and it came in a nice metal storage box as do most Dumore grinders. But if you will make for yourself a place to mount the grinder somewhere in your shop, on the wall or on a workbench or even on a stump if you have one high enough, you will find that you will use it nearly every day.

Photo 22. Below, the same dresser used on another grinder. Adapt the dresser to as many of your grinders as you can. This makes it even more useful.

Photo 23. Bottom, tool post grinder mounted on work bench.

Photo 23 is of the place I have mounted mine and you can see in the picture some of the assortment of wheels I keep handy. Install it so that it is well lighted, put in a work light just for this purpose if necessary, and put it into a handy and comfortable position. Mine is about shirt pocket high and is within four feet of both of my lathes. Throwaway inserts are great, no doubt about it, but sometimes there is no substitute for the hand ground cutting tool and this little grinding machine will be right there when you need it.

These six little projects, if you do them all at the same time, won't take you more than a day and a half to complete. You can spend more time on them but the basic tools are not complicated. And they will save you hours and hours of time over the span of just a few years. There are many, many more. But the best of ideas isn't worth much to you until you take the time to make the tool or accessory and put it to work in your shop.

Chapter 12

The Engraver's Vise

The Engraver's Vise

The engraver's vise is known by several different names. Engraver's vise, engraver's ball, ball vise and engraver's block are just a few. Regardless of what it is called it is a marvelous piece of furniture for the top of your work bench and it is a great machining project. If you have one of these vises on your bench you will be surprised at the number of times you will find a use for it. And you may even use it for engraving! Photo 1 is a picture of my newest and latest vise, made especially for this chapter. If you are interested in making one of these vises, or <u>d</u>evices as the case may be, there are complete plans and drawings for doing just that a little further on. But before we get to that here is a little more information about the project.

Photo 1. The vise described in the plans and drawings given here.

Photo 2 is a picture of a ball vise I made over 30 years ago. Many things have changed for me since that time including the variety of materials in my "leftovers bin" and the machinery I have in my shop. But I apparently did a pretty good job making the vise because it shows no sign of wear after 30 years of use. Intermittent use, I will admit, but it doesn't sit neglected all the time. There are some major differences between this vise and the new one which may influence how you go about making yours. We'll discuss that some more a little later as well.

Photo 3 is a different approach to the engraver's vise. I visited with Ray Viramontez in Dayton, Ohio a couple of decades back and this vise is built according to my understanding of his design. Ray is retired from the U.S. Air Force and is an accomplished firearms engraver. This vise turns freely but only in one axis whereas the ball vise is free to rotate as well as to nutate. The axial bearing in this vise may serve as a bearing in the vise you build instead of the tapered roller bearing shown in the plans.

Photo 2. The ball vise I made about 30 years ago. I treated this vise with a cold bluing solution which has held up remarkably well over the years.

Here are more alternatives you may wish to consider in building a vise.

The Engraver's Vise

Scaling down some. The vise in the plans included here finished up at a little over 43 pounds. Weight is a good thing because it makes a steady platform to work on but a 5" ball, or even a 4" ball, will still be substantial and may fit your methods and materials better.

The T-slot on the top surface of the vise and the matching T on the bottom of the jaws are machined from solid material in the plans presented here. When I made the first ball vise I had neither a T-slot cutter nor Woodruff cutters so I made the slot and matching jaws from multiple parts. Photo 5 will show you what I am talking about. It worked just as well and was actually a little easier to make. If I ever make another one I will probably go back to that method.

Photo 3. Ray Viramontez' plan for an engraver's vise. Note that one of the top jaws has the ability to swivel. This is handy for holding irregular shaped parts for engraving.

The bearings used in this vise are strictly a function of what was in my used bearings drawer when I started the project. (The only time I will ever throw away a bearing is when it is completely frozen up and even then I have been known to use them for inert spacers.) A large tapered roller bearing is convenient and works well but anything which will take a slight thrust and still rotate freely will work. The only thrust involved is the weight of the vise spindle, the vise jaws and whatever work piece you may have in the vise. Photo 4 is the bearing in the Viramontez vise we looked at earlier. It was machined from bronze and the lubrication grooves in it hold the grease so that it has never required additional attention. Using this type of bearing has the added advantage of eliminating the need for a lower bearing.

Another difference between the vise in the plans and the first ball vise I made is shown in Photo 6. I used a different method for locking the vise against rotation in the second vise. Either method shown will work but I do think the method included here in the plans is the best. The photograph shows the method used in the old vise. It consists of a tapered bearing retaining nut on the spindle with a brass plug bearing against the tapered surface. On the second vise there is no retaining nut provided for the bearing. Since all the forces are downward, and they are pretty minimal as forces go, gravity does a great job of bearing retention. The purists among you will probably add a locknut and that is OK with me.

And one final difference between the two projects. On the original vise I applied a treatment of cold bluing solution called Oxpho-

There are three classes of broken. Class 1. Broken but I can have it fixed for you in just a little while. Class 2. Broken and I can fix it but it will take a while. Class 3. You better get started looking for a new one.

The Engraver's Vise

Photo 4. The bearing for the Viramontez vise. This bearing may be substituted for the ball vise and will eliminate the need for a lower spindle bearing. I lubricated this bearing when I made it nearly 30 years ago and it has never required further attention.

Blue™ from Brownell's, Inc. It has held up remarkably well over the years and has, I am sure, prevented the vise from rusting. I have not applied any treatment to the new vise yet but I probably will.

If you choose to make this vise there are a couple of things for you to consider before you begin. I have described the process of machining the spherical profile of the ball itself in Chapter 17 on making a profiling attachment for your lathe so I won't repeat that here. Drilling the hexagonal hole into the end of the vise screw is also described in another place, Chapter 19 on Watts tooling. We will talk some about alternatives to drilling a hexagonal hole however. Alternatives are what makes the world go around in my shop and the fun of a project is often made greater by finding different ways of doing things. If you know me at all you will have heard me say that the real craftsman is not the man who never makes a mistake but rather the man who can make a mistake and devise some way of correcting it without losing the work already invested in the job. That way you may get out of having to tell about your mistakes! You will notice in this chapter that I have not admitted to making any mistakes. If you would like to go on thinking that, it is OK with me.

The final caution I will mention is this. Don't drop this sucker on your foot! The slick round surface of this ball is amazingly difficult to hold on to and it can really ruin your day if you

Photo 5. An alternative method for machining the T-slot and matching vise jaws for the top of the ball vise. This worked well when I made the original vise and is perhaps a little easier than machining the T and T-slot from solid material.

let it get away from you.

Other options are always available and should be considered but keep in mind the purpose of the finished vise and don't make substitutes which will affect the completed job. You can, for example, use one fixed jaw and one movable jaw. You can use welded components for the jaws or even for the ball. Stacking rings cut from hot rolled plate is an example of material substitution. Stack the rings, weld them on the inside and machine the ball from the weldment. You can, if you choose to go the other way, machine all of the components from a high strength steel alloy and have it chrome plated but there are limits. The only thing that is not limited is the imagination of the person doing the job and you will probably come up with options I have not only failed to mention but would have never thought of.

If you do decide to build the vise described here with these drawings and photographs then this is the way I did it. If you should decide to build one similar but different this may serve as a guide. If you decide not to build one at all you are welcome to come to my shop and use one of mine. I now have two of them.

Photo 6. Another design change. The method for locking the vise against rotation in the original vise was a brass plug as shown here. It was tightened with a ¼" hex wrench from the outside of the ball. The method described in these plans requires a little more effort but it works much better.

The Engraver's Vise

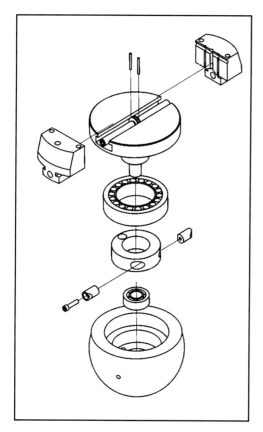

Drawing One. An exploded view of the vise and all of its components. This shows the relationship of all the parts to all of the other parts.

Bill of Materials, Engraver's Vise

- 6.5" diameter CRS steel. A total length of approximately 12" will be required to machine all three major components. The ball, the spindle and the jaws. If you buy new material buy 12L14 steel. The difference in machining characteristics will more than offset the difference in the cost.

- 3.5" diameter steel. A piece 1" long for making the locking ring.

- 5/8" diameter CRS 2-1/2" long for clamping pins.

- 1" diameter CRS 7-5/8" long for the vise screw.

- Timken cone assembly #395A

- Timken cup #394XS

- Ball bearing #204. A sealed bearing is good here but not really necessary.

- (2) 1/8" roll pins 1-1/4" long.

- ¼-20 Socket head cap screw 1" long.

- 3/8 16 Socket head cap screw 1-1/2" long

Photo 7. Boring the internal diameters in the body of the ball.

152   Building Shop

Begin the job by sawing a piece of the 6-1/2" CRS to 4-3/4" in length and chucking it in the lathe. Face the part, bore the internal profile to its finished dimensions and remove it from the machine. Although it is not impossible to chuck the part after machining the spherical profile on the outside it is decidedly easier to do it this way. See Photo 7 and Drawing Two.

Before you go to the milling machine and set up for machining the hex wrench access hole, machine the locking collar, Drawing Four, and install it in the body of the vise. (Note: Leave a little material in the bore of the locking collar for the later machining of the locking pins.) Drill and tap the 3/8"-16 hole and bolt the collar in place. Carefully locate the work piece relative to the spindle and drill through one side of the vise body and through the locking collar with a 17/64" bit. Use an end mill to start the hole in order to overcome any tendency the drill may have to wander because of the curved surface of the vise body. After removing the locking collar from the vise body set it up separately in a vise and, using the 17/64" hole as a locating point, drill through with a 5/8" diameter drill. If you choose you may continue at this point and complete the machining

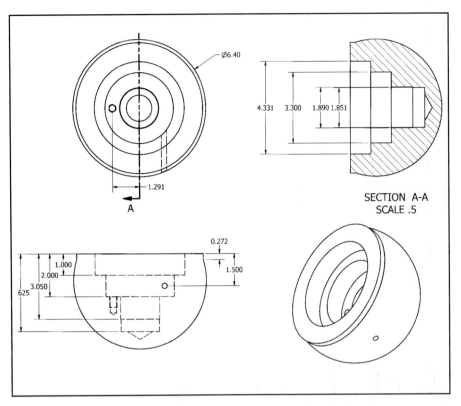

Drawing Two. The body or ball.

Photo 8. Roughing out the spindle. Love that 12L14 steel!

The Engraver's Vise

Photo 9. Above. Setup for boring the locking collar with the locking pins in place.

Photo 10. Right. Locking collar and pins completed and located in the body of the ball.

of the collar and its locking pins. The locking pins are simple pieces of 5/8" diameter CRS with a clearance hole and counter bore in one and a ¼"-20 thread in the other. Refer to the photographs and to Drawing Seven for this process.

As we have already mentioned, the contouring attachment described in Chapter 17 covers the process for machining the outside contour of the ball. I will say here that this is a great opportunity for you to use your own method for machining spherical contours on an engine

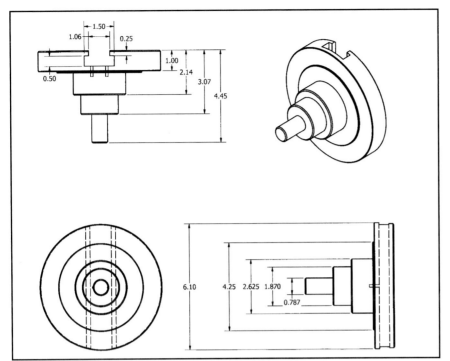

Drawing Three. The spindle and top surface.

lathe. There seems to be a number of options, most of them good. I have seen over the years many different approaches to this old procedure and some of them get pretty complicated. It is a fun thing to invent and it has been invented many, many times. The best one may still be waiting to be invented.

Now let's go to the spindle. Sequencing here is largely a matter of choice. In this example I began by facing the material and turning the finished diameter of the top of the spindle. I then went to the milling machine and machined the T-slot before I turned the diameters for the bearing and locking collar. It is six of one and a half dozen of the other and really makes no difference. There is nothing difficult nor unusual here in machining the diameters for the bearing fits. Don't try for press fits on the

Drawing Four. The locking collar.

Building Shop

The Engraver's Vise

Photo 11. Above. The first step in machining the T-slot.

Photo 12. Right. Machining the T. The cutter here is a T-slot cutter but Woodruff cutters will work. Just a little slower.

bearings because the loading doesn't justify it. Just make a good push fit for both the upper and lower bearings to be assembled to the finished spindle. Whether you machine the T-slot first or last you will set the part up as shown in the photographs and machine the slot in the top face. This is a case of "whoever gets there first". By machining the T-slot before you make the vise jaws you make it easier because you don't have to pay such close attention to sizes and

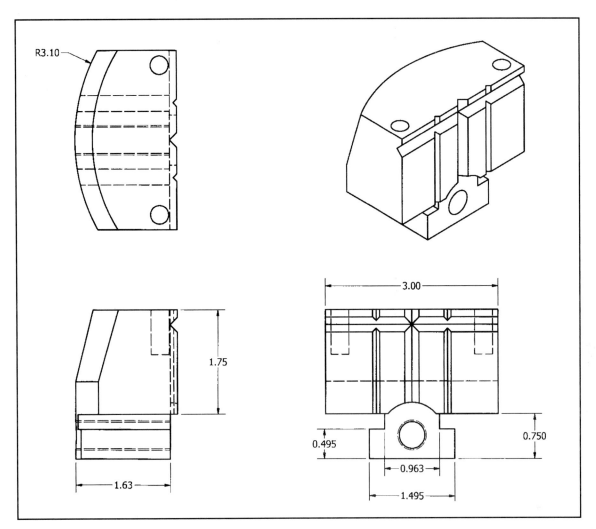

tolerances. Machine the slot pretty close to the dimensions given in the drawings but if you miss the size you can make the T of the jaws to fit. A good finish is a more important requirement here than an exact size. Before removing the part from the mill, drill the two 1/8" holes in the top for the roll pins and the spindle is complete.

On to the vise jaws. They will be machined as one piece almost until they are finished. Set the material up in the lathe and machine the tapered top of the jaws. The taper is mostly cosmetic but that doesn't make it a bad thing. It needs to look good. After machining the outside diameters and facing the top of the jaws, go to the milling machine and machine the T to fit the slot on the top of the spindle. What you are looking for here is a smooth sliding fit with no interference. You will not be able to try it until after you have cut away the extra material on either side of the T so now is when you need to pay attention to your clearances and dimensions. I have found that if you allow from .003" to .005" on all sides you will arrive at a pretty nice fit. A little too much is better here than not enough.

Drawing Five. One of the top jaws. Although two jaws are required, only one is shown in this drawing. Remember that one jaw has right hand threads and the other has left hand threads. The holes and V-grooves may be added as needed to accommodate whatever you may be working on.

The Engraver's Vise

Photo 13. Finishing touches. Machine a chamfer on the top edges. A file will work for removing the burrs from the bottom edges of the slot.

After sawing away the bulk of the unwanted material on the sides of the jaws, machine the sides to be parallel with the T and with each other. This is important because it will affect the alignment of the hole you will drill for the screw. Use the same side of each jaw to set up for drilling. The hole must be parallel to the T or otherwise there will be a binding with the screw in place. Tap one jaw for ½"-20 RH and the other for ½"-20 LH. It makes no difference which one is which. Refer to the photographs for a more complete understanding of the process. The V-grooves in the faces of the jaws should be strictly to meet your own requirements. You may not want them at all. And if your requirements change it is no problem to add the V-grooves as well as the holes in the top of the jaws. These should almost be considered to be in the same category as soft jaws and making a spare set can be justified.

Photo 14. Machining the bottom sides of the vise jaws. This is easier if you machine the T of the jaws before removing the extra material. It gives you a place to clamp the job to the table.

158 Building Shop

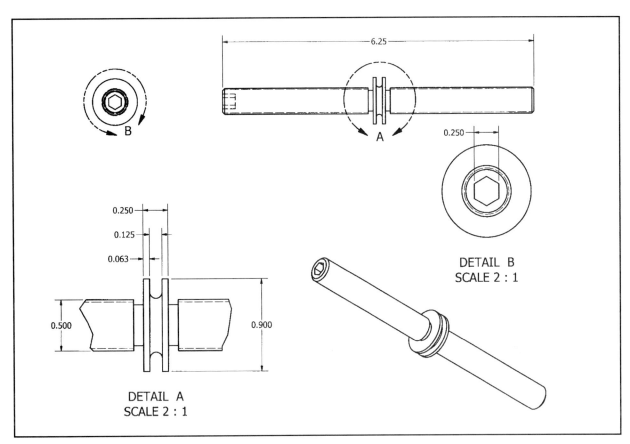

Drawing Six. The screw. One end left hand threads and the opposite end right hand threads.

Machine the screw for the vise as shown in Drawing Six. If you do not have access to Watts drills and tooling there are several alternatives. You can leave a little extra length on one or both ends of the screw and machine an external square or hex for using a wrench. You can drill and tap for a 3/8"-24 setscrew in one or both ends and either jam the screw into the bottom of the hole or LocTite™ it into place. It you choose the setscrew method you will have to use a smaller hex wrench but that is no real problem. Make sure the width of the groove in the center is large enough for it not to bind on the roll pins or dowel pins. There is no real force applied to these pins. They are merely to keep the jaws centered on the top of the vise. The pressure is applied to the jaws themselves.

Photo 15. Sawing away the extra material.

The Engraver's Vise

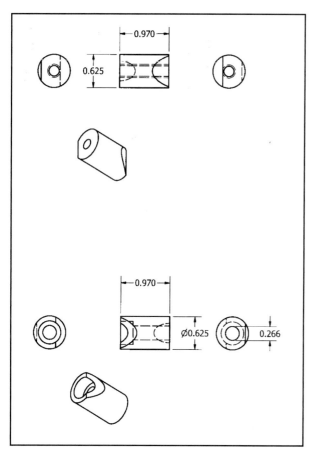

Drawing Seven. The locking pins. Refer to the text for machining instructions. These pins should be machined at assembly with the locking collar.

And that is pretty much all there is to it. Assemble all of the parts and check for smooth movement and good fits. The jaws should open all the way out without any roughness or binding. The top should turn freely and without any drag until you lock it against rotation and then it should not turn at all without turning the ball in its bed. The photographs show a couple of different resting beds for the vise. I did not make one of wood but I have seen some nice ones made of hardwood and they can be fun to make as well. This engraver's vise can be a great project and opportunity for you to make something else to have in your shop. And you might even use it. As a matter of fact, you might even use it for engraving! Wouldn't that be something!?

Photo 16. Machining the sides of the vise jaws. The top of the milling machine vise serves here as a way of keeping the surfaces of the jaws parallel with each other and with the T.

160   Building Shop

The Engraver's Vise

Photo 17. Above. I had already machined this jaw when I remembered to take the photo. But here is the setup for machining the face, drilling and tapping the hole, and machining the relief for the center of the screw.

Photo 18. Left. The completed ball and spindle minus the top jaws.

Building Shop 161

The Engraver's Vise

Photo 19. The finished project. Notice the large capacity of the opened vise.

Chapter 13

A Welding Quandary

Welding

I am frequently inspired to respond to questions and comments I see on the various web based bulletin boards found on the Internet but I find that I am so long winded sometimes that I cannot provide a short and succinct answer to a question, even though I may know the exact answer to the question being asked. So I usually just go on and read the next contribution to the board and forego the opportunity to share with the metal working world my vast store of wisdom (AHEM).

But sometimes I will see a really good and pertinent question, one that my particular slice of experience can address, and since others in my past have always been willing and generous in sharing their lessons with me I will try to answer the question. A case in point was a question asked on the BBS sponsored by Village Press in November of 2007 by a contributor who goes by the screen name of BIGBOY1. I have asked his permission to quote him here and this is his question as it appeared on the BBS.

Welding Quandary

I'm in a welding quandary as I'm debating on the type of welder to buy. I'm looking at either a MIG or TIG type welder but I would like to hear opinions from the group. My requirements are to weld both steel and aluminum, with the thickness being ½" or less. At this point in time, I'm leaning towards the TIG machine for the following reasons: Changing the weld wire must be done to change between steel and aluminum welding. To do so, the MIG machine has to be opened up and the steel spool of weld wire replaced with a spool of wire for aluminum welding. To change welding materials on the TIG machine, different weld rods are chosen. With a TIG machine, stick welding can be done in addition with some machines while it is not possible with a MIG machine. Some TIG machines have the feature where the tip is placed where the weld is to be made and when the handle is lifted, the arc starts. This prevents 'wandering all over the part' with the strike of the arc when brought in close contact.

Would appreciate any thoughts or ideas.

Bill

There were at the last time I checked, more than four pages of responses to this question and nearly every one of them was helpful and useful. Suggestions and experience-based information abounded as professional welders and welding machine sales/service people offered their solutions to the quandary. As you

---

I look at some of the things men made before the day of the modern machine tool and I cannot help but wonder: What could those guys have done with that talent and the tools of today?

would expect, there were also several endorsements and several condemnations of the different makers of welding machines.

So here is what I am going to try to do in this chapter. I am going to try to offer some hopefully useful information for the hobbyist, the novice, the wannabe welder or just to the interested bystander about selecting which process, and consequently what equipment, is best for your purposes. But first, here are all of the reasons why you should ignore what I am saying altogether.

- I am not going to endorse, I am not even going to mention, any brand names of welding machinery other than to say that I have welded with Airco, Esab, Forney, GE, Hobart, Lincoln, Marquette, Miller, P&H and Westinghouse welding machines. (My apologies to the ones I forgot about.) I know some of these brands are made by the same people; Miller made Airco machines for a while back in the 1960's, for example. They all did the job they were called on to do at the time although some did it better than others.

- I am not knowledgeable about the electrical terms used to describe some of the many variations of welding machines. I don't know a whole lot about what goes on inside of that cabinet so don't be too hard on me if I call something by its wrong name. Most of my knowledge and experience has its source out at the end of the welding cable.

- The advice I will give here is free and we all know what free advice is worth.

- The advice here is also based on my experience and my opinions. Only I know what that is worth.

- And finally, only you can really answer the question about what equipment you need because only you know what your requirements are. What I am going to try to do here is provide some guidelines to go by.

Safety First

Always a good rule and a great place to start. Arc welding is inherently a dangerous occupation. You are dealing with a molten puddle of metal which has temperatures in excess of 2000 degrees F, relatively high electric currents, high concentrations of fumes and possibly toxic smoke as well as extreme concentrations of ultra violet and infra-red emissions. These are the direct dangers. Indirectly you may be on a ladder or a scaffold or under a truck frame or, and here I can attest first hand, being lowered by two men holding on to your heels into a concrete form to weld something that somebody forgot to weld when it was in the open. And yet, with all of the potential hazards present when arc welding, it is one of the most popular home shop occupations, next to wood turning, to be found in our basements and garages. We are either doing something right or we are luckier than we deserve!

Your safety equipment will consist at a minimum of:

- An approved welding helmet or hood. And by approved I don't mean just that your wife likes the color of the design you have painted on it. It must shield all the exposed skin of your face and neck. I use a No. 10 filter lens for most of my welding but if you are using higher amperages, as in arc gouging for example, or if your eyes are particularly sensitive to light, you may want to go as high as a No. 14. If you have doubts about what you require consult with an optometrist or other qualified eye specialist and get a professional opinion. But don't set in to weld wearing just a pair of aviator shades regardless of how "cool" they look. The auto darkening lens is a good thing. I didn't think so when they were first introduced but I use one now and I wouldn't like to have to give it up. As you get older you will also find that you require a magnifying lens or "cheater" lens in your hood.
- Safety glasses should be worn under the welding hood. The welding hood itself gives pretty good eye protection while it is down. But when you raise the hood your eyes are vulnerable. A good pair of safety glasses worn under the hood is not only a good idea it is a common sense requirement.
- Welding gloves or gauntlets. I will frequently weld without gloves but I don't claim it to be a good idea. Gloves should be worn, leather not cloth, and they should cover the skin at your wrists. If you go without gloves too often you will get a burn. Guaranteed.

Other safety equipment will depend to a great extent upon the job. If you are doing some serious overhead welding, 5/32" 7018 electrodes on the bottom side of a heavy wall pipe for example, a leather jacket or leather sleeves will be appreciated. As a general foreman I once knew said, "When you are welding vertical or overhead with 5/32 you will lose a tablespoonful ever once in a while". You want it to splash off instead of sticking to something.

As we learn more about our surroundings we learn more about what can hurt us and welding fumes have now made that list. I did most of my serious welding back in the 1960's and 1970's and we ignored, the entire industry ignored, at that time many dangers because we just didn't know they were dangers but that has changed. Or at least it is changing. Wearing a good filter mask now can be defended when back then it was considered "sissy". Masks can be passive, just a simple filter worn over the nose and mouth, or may require a purified air supply, depending on the application. If you are hired to weld on a pipeline at a hydro-electric plant which was built in 1940 the chances are good there is some lead in the paint and you don't want to mess around with lead.

To sum up safety in welding here are some good rules:

- Don't allow any of yourself or your skin to be directly exposed to the welding arc. Even for brief periods.
- Don't weld for extended periods in a confined space.
- Don't weld near any flammable substances. Keep a fire extinguisher handy wherever you are welding.
- Don't weld under any sort of suspended load without shoring up.
- Always place the ground clamp as near as is practical to where you are welding. Never allow welding current to pass through gearing or bearings.
- If you have any doubts about the safety of the materials or conditions you are asked to work with or in ---- question it. If the boss says do it and you know it has preventable risks associated, your best course of action is to be looking for a new job!

Why would you spend hours and hours making something you can buy for a few dollars? I guess you just have to be there. It is like assembling a jigsaw puzzle instead of buying a picture.

Now, with that out of the way, let's talk about the processes and what may be expected of them. Comparing welding processes, MIG to TIG for example, is like comparing the Green Bay Packers to the Los Angeles Dodgers. There are just too many basic differences. So I am going to list the ones I know about and have had experience with and then list their strengths and their weaknesses. Let's begin with stick welding.

Stick Welding

Stick welding, or SMAW (Shielded Metallic Arc Welding), as it is now referred to, is where most of the welders of my generation started out. There are many choices of machine variations available for this method and here are a few of them:

- The "buzz-box." This small, transformer-based welding machine can be found by the thousands on farms and in garages all over the country. It provides, usually, up to 225

Amps of welding current in AC only. It will not produce 225 Amps for very long at a time but for welding with 1/8" electrodes where nothing much more is required than to stick it back together, these machines can be very serviceable. When combined with a high-frequency unit they also serve as a good power source for welding aluminum, zinc die cast or magnesium using a TIG torch. These machines typically require only a 50Amp, 220Volt, single phase power connection, the same as most kitchen stoves.

- The motor-generator, or MG, machine. I don't see as many of these machines in shops as I once did. They made for serviceable machines and consisted of an electric motor driving an electric generator. The terms "coffee pot" and "torpedo" were assigned to them depending if they were of vertical or horizontal construction. They are capable of fairly high sustained amperages and perform excellently for arc gouging operations. Most of the ones I have had experience with required 3 phase power as their primary connections. The diesel or gasoline powered portable welding machines in use today are all variations of the MG machine but I have not owned nor used a portable machine in about twenty years now and I like it that way. If it can't be brought into my shop and put on the welding table, chances are I will not take the job.
- The constant-current, AC/DC welding power source. This is the machine I want to own if I am only allowed to own one welding machine. They are available in any output configuration you could possibly need, 250 Amps is sufficient for most shops, and all of the "bells and whistles" accessories are there as well. I will talk about those later. These machines usually will operate from a 220 Volt, single phase power source.

Tig Welding

Next up the progression of welding processes is TIG (Tungsten Inert Gas) welding. It too, has a newer acronym, GTAW (Gas Tungsten Arc Welding). When I first learned TIG welding it was known as Heliarc welding because it used Helium gas as an arc shielding gas. The only thing required for TIG welding in your shop is a suitable torch, a bottle of gas, usually Argon, and a regulator. And a suitable power source, of course. TIG welding is done using either AC welding current or DCSP (Direct Current Straight Polarity) meaning that the electrode is negative. TIG torches are available either as wet or dry, this designation meaning that they are either water cooled or air cooled. Water cooled torches are typically of lighter construction than air cooled and I can tell you from firsthand experience that you don't want to try welding with a water cooled torch without the coolant flowing. TIG welding is slower than stick welding but if you need to do delicate welding, or if you are welding alloys other than mild steel, then it has no equal.

Mig Welding

In BIGBOY1's bulletin board question, he is looking for advice as to whether or not to purchase MIG or TIG for welding both aluminum and steel. Either process will do either job but not equally well, or at least not equally as conveniently. To weld aluminum with the MIG process you must have, or at least it works better if you do have, a spool gun. I bought a MIG machine which was being introduced by the manufacturer as a new machine and as an introductory offer the spool gun was included. I probably would not have purchased it otherwise. But here is why you need a spool gun when welding aluminum. The aluminum welding wire I use on my machine is .030" in diameter. If you were to try to push that size aluminum wire through a 10' long feed tube the likely result will be a ball of aluminum and a badly frayed temper. The spool gun allows the wire to be pushed only through about a 6" length of feed tube before being deposited as weld metal. And remember, welding aluminum with a MIG requires Argon gas rather than the CO2/Argon mix usually used when welding steel. I use my MIG machine more than I do my stick/TIG machine but I will still stick to what I have said. If I can only have one machine it will be the stick/TIG machine.

There are other options. FCAW (Flux Cored Arc Welding) is simply MIG welding without a gas bottle. That probably is an over simplification but the shielding gas is not needed with the flux cored welding wire. I have little experience with it so I won't talk too much about it. And I am intentionally leaving out of this discussion the options of O/A (Oxygen/Acetylene) welding and brazing or soldering. All good and acceptable methods for joining materials but not really the subject here. Maybe some other time.

And now the promised consideration of the "bells and whistles" available in welding machines. Referring again to BIGBOY1's posed question, he mentions, "Some TIG machines have the feature where the tip is placed where the weld is to be made and when the handle is lifted, the arc starts". This is called lift-start by the machine builders and, although I have never used it, I do see where it might be handy. I have found that the use of the self-darkening welding hood makes that option unnecessary. Others may disagree. Other options available on today's welding machines include, but are not limited to:

- Current decay. An option which allows you to gradually reduce the current at the end of the weld to avoid leaving a crater.
- Post flow gas timer. Allows you to leave the shielding gas on the weld after the arc has stopped until the puddle has had time to solidify.
- Pre flow gas timer. An option to allow the shielding gas to completely cover the area to be welded before the arc is initiated.
- Arc control for stick welding. This gives some variations in the arc characteristics when welding in the vertical or overhead positions.
- AC balance. An adjustment to provide either deeper penetration or greater cleaning action of the welding arc.

There are others, probably many others but if I get too involved in an in-depth discussion of welding options and features available on today's machine, I will reveal myself as the antique I am rapidly becoming. And now, after wandering all over the welding table and providing information of marginal usefulness, here is my official answer(s) to BIGBOY1's question.

1. Best option — Buy a TIG/stick welding machine. For the basement/garage shop the capabilities of this machine will best fit the needs of most weekend welders.

2. Take aluminum out of the question and I would say buy a MIG. These machines are available in 110V configuration but if you have the power and the money to buy a larger, 220V machine, I suggest you stay away from the little guys. A MIG gives you the capability to weld practically any steel job that comes up and they are fast and handy.

3. For Heaven's Sake buy some sort of welding machine! I can't believe how many young couples there are who get married and try to set up housekeeping without even the most basic of welding equipment. I don't see how they make it!

As I said at the beginning, only you know what your planned projects are. If you know that you are going to be fabricating ornamental iron projects and nothing else then the answer becomes simple. If you are going to try to serve the public as a neighborhood fix-it guy then you will need every type of welding machine known to modern man and a few that haven't been invented yet. Maybe we should think of welding machines as I do swimming pools. I don't want to own one but I sure would like to live next door to someone who does.

If you are not in the habit of wearing a glove on the hand holding the part you are de-burring with a file you are going to de-burr some knuckles.

Chapter 14

Turning Between Centers

Turning Between Centers

LBS (Long Boring Story) coming up so bear with me. I was a Machinery Repairman Third Class Petty Officer in the Navy when my ship went into the Portsmouth Naval Shipyard for an overhaul. The crew lived aboard during the six-week process and I, along with a couple of ship fitters, found that the shipyard was so lacking in workers they were willing to hire us on an hourly basis when we were off duty. It was a good deal for us. We made more wages than any of us had ever heard of, I think we were paid something on the order of three bucks an hour, (this was in 1960) and it was a great place to escape from the constant chipping and welding that was going on in our living quarters. The arrangement didn't last long but that is another story.

On my first day as a hired machinist in a real machine shop I was assigned the job of making some sheave pins for crane pulleys. I was told where to go to get the material and where the lathe, a 20" American Pacemaker, was located. I went happily off to get the 3" diameter piece of 4140 steel expecting to be given a piece long enough to chuck one end while machining the required diameters on the part. The stockroom attendant sawed my material to exactly the length of the finished part! There was just enough extra length of the material to take light facing cuts on either end. How am I going to hold on to this thing to machine the full length? The answer was, of course, to set the part up so that I could turn it between centers. I have turned many parts since then using this method and it is considered, at least by me, to be one of the basic processes for a machinist to learn. Here is one way of doing it.

Photo 1. A combination drill and countersink used for machining centers in a part.

First cut your material to length using the same stingy process the stockroom attendant at the shipyard in my story did. Saw to length plus about 1/16" inch. If you are sure your saw will produce a square cut you can leave a little less. All that is required is that you have enough material to face the ends clean and stay within your length tolerance. Next machine center holes in both ends. Photo 1 is a picture of the combined drill and countersink used to produce the center hole. It is important that the center hole be:

- In the center of the part
- Deep enough to insure it won't come out when cutting force is applied
- Have a smaller diameter hole in the center so that the point of your lathe center will not be in contact with the bottom of the hole
- Clean and free of chips and debris which might cause the part to run out

Photo 2. Drilling the center hole in one end of the material.

Photo 2 shows the center hole as drilled in one end of the material. If the job is too long to support securely in a chuck and if your lathe does not have a hole through the spindle large enough to accommodate it, you have a couple of choices. You can either use a steady rest to support the work while drilling the centers or you can locate and center punch the locations or drill them with a hand drill. I have done both and both methods work. Facing the part gives you a nice finished end and it also provides a clean and square surface in which to drill your center hole.

After the work piece is prepared you must prepare the machine. When this is done you will have:

- A live center in the headstock which will accept the center hole you have in the part and which is running dead true with the axis of the lathe.
- Some means of driving the part such as a face plate, a drive plate or a chuck with jaws which will engage and drive the lathe dog.

Photo 3. A center for the headstock. This must be re-machined every time it is removed and replaced in the chuck.

Turning Between Centers

Photo 4. Setup for machining a 60 degree center.

Photo 5. Machining the center. The tool must be hand fed using the compound rest.

- A tail center which will accept the center hole in the other end of the part and which is in perfect alignment with the center in the headstock. This may be either a ball bearing live center or a hardened center with adequate lubrication.

One method of preparing the headstock center is shown here. First, make sure you are finished with the process of preparing the material. Once you machine the center in the headstock you should not remove it until the job is done. If you do have to remove it for any reason you must re-machine the center when you replace it. The accuracy of your job will depend to a great deal upon the accuracy of these centers. When you are ready for the center, mount

Turning Between Centers

Photo 6. A sampling of bent tail lathe dogs.

an appropriate piece of material, see Photo 3, in either a three or four jaw chuck. Note that the material has a shoulder which bears against the front of the chuck jaws. This will insure that the center will not be forced back into the chuck under the pressure of the cut. Set the compound at 30 degrees from the axis of the machine, see Photo 4, and machine the point of the center as shown in Photo 5. Many lathes, particularly the smaller bench or toolroom lathes, are equipped with a center which is inserted into the headstock of the machine. This works well but I prefer the method shown here. With this method you can control the length the center protrudes from the chuck which is sometimes an advantage. You can also be sure that the center is running in perfect concentricity with the axis of the lathe and you can use one of the chuck jaws to drive the lathe dog. Whatever works for you is the method to use.

Now mount the lathe dog on the work, Photo 6 is a picture of a small collection of dogs I keep handy, and place the part between the centers as shown in Photo 7. I am using here a ball bearing live center in the tail stock which, in my opinion, is the best choice for this type of work. Dead centers are typically more accurate and are widely used in grinding operations but the loss in accuracy in lathe work is minimal and is, again in my opinion, more than offset by the convenience of not having to keep the centers lubed.

Spending an hour and a half removing the remains of a $7.45 tap from a $350 part because you didn't want to throw the tap away is not what I call being frugal.

Turning Between Centers

Photo 7. Above, a part set up between centers.

Photo 8. Right, test cuts to determine the amount of taper — if any.

Now on to the last preparatory operation of setting up work to turn between centers, which is insuring that the job is running true and without taper. If the tail center is offset at all you will machine a tapered part. In fact, this is a good and accepted method for producing tapers but that is a subject for another time. You want

Photo 9. Measuring for taper.

the part we are machining to be straight and here is how you can make it straight. After you have checked all of the variables —

- The dog screw is tight, (Watch out for that dog. It will bite if you get your elbow into it!)
- The dog tail is firmly engaged against the chuck jaw or drive plate slot
- The tail center is tightened so that there is no slack in the setup
- You have checked that the compound will not try to occupy the same space as the lathe dog

—you are ready to take your first cut.

Begin by machining a short diameter at the tail end of the work. Set the cross slide or the DRO to zero and back out. Then move to the other end as shown in Photo 8 and machine another diameter at the same cross slide setting. Measure the two diameters, Photo 9, and note the difference. If there is no difference, or if the difference is within the tolerances you are given for taper, proceed with the job. If there is a difference you must make some adjustments.

For the sake of this instruction let us say that the diameter of the work at the tail end is .004" larger than the diameter at

> "That tool has nearly 4 inches to go before it reaches the chuck. I have plenty of time to go fill my coffee cup." Yeah! Right!

Turning Between Centers

Photo 10. Adjusting to remove taper.

Keep in mind that if you have to find the time to do the job over you had the time to do it right in the first place.

the other end. You must move the tail end .002" closer to the cutting tool in order to make the two ends the same. Set up a dial indicator, see Photo 10, so that it bears against the machined diameter on the tail end. You must then loosen the clamp which clamps the tail stock to the lathe bed and, by tightening and loosening the screws at the base of the tail stock, move the work piece the proper amount in the proper direction.

The hex wrench in the photo is what adjusts tailstock setover on my lathe. Yours may be different but there will be some method provided for this adjustment to be made. After you get the desired reading on the dial, tighten everything back down again and repeat the process by taking another light cut on the diameters at each end. Make any further necessary adjustments and you are all set.

Most any machining process, particularly shaft work, which can be done in a chuck may be done between centers. But as in everything you do in the machine shop — pay attention to what you are doing. For example, if you are threading and you want to remove the part from the lathe to check a fit, be sure you put it back with the dog having the same orientation to the driving surface.

And this bears repeating. That lathe dog will bite you if you let it!

178   Building Shop

Chapter 15

Machining with Soft Jaws

Soft Jaws

Here is one of those, "Betcha didn't know," things. Betcha didn't know that a good bird dog and an engine lathe have something in common. They both are better if they come with soft jaws. That is probably a stretch for coming up with an attention getting first line but there is truth to it.

I am, of course, talking about the soft jaws used on a three jaw chuck. They are also used on six-jaw chucks and most any other chuck which comes equipped with two-part jaws. Those two parts being a top part which is hardened and usually reversible and the bottom part which engages the scroll of a universal chuck. Photo 1 shows the type of chuck jaw being referred to. The top jaw may be removed and either reversed for chucking larger diameters or it may be replaced with a soft jaw of your choosing. Here are some of the advantages of using soft jaws on your lathe:

You only are fooling yourself if you think you can save money by buying cheap tools.

- They are, as the name implies, soft. They can readily be machined to fit your part and may be used in operations where the finish on your part needs to be protected from what I call "chuck trauma."

- They are useful where multiple parts, or multiple operations on the same part, are involved. They offer a repeatability which is usually not available even on the most accurate of standard chucks and chucking setups.

- They can be configured to accommodate a large variety of shapes. I use them frequently where the parts being machined are thin walled and require a firm yet tender chucking pressure.

- Soft jaws may be welded to aid in accommodating some of the less usual shapes you may encounter in lathe operations.

Photo 1. The 2 part chuck jaw. The top jaw is removed to install the soft jaw.

Soft jaws are a necessary part of my shop operation but there are some facets to their use which are sometimes either not known about or are ignored. I am going to talk a little about some of them here.

Photo 2 is a picture of three chuck jaws, one each from three different sets of jaws which I have in my lathe cabinet. I have probably 20 different sets of jaws in that cabinet and will probably never use any of them again exactly as configured.

Soft Jaws

But I can frequently find a set which will serve with only a minor modification. That is why I keep so many of them on hand. A large variety of soft jaws are available from many tooling specialists in either steel or aluminum. I use steel for the most part because it is so easily welded and many times welding an extension to the front of the jaw helps in achieving exactly the correct profile.

One of the most important, if not the most important, thing to remember when machining a set of chuck jaws is that pressure must be applied in the same direction as the pressure you will apply when machining your part. That sounds confusing, doesn't it? Think of it this way. If you are going to chuck the OD of a part in soft jaws, you must machine the jaws while the chuck is clamping the OD of a sample part. Photos 3 and 4 might help. In Photo 3 you see a stack of circles I keep in my chuck drawer cabinet just for the purpose of having a clamping sample for many different sizes. In Photo 4 you see one of these circles clamped in the chuck with the boring bar in place preparing to bore to the correct size for the job. Photos 5 and 6 are of the same things except these are for turning jaws to accept the ID of a part. The thing to remember is that the jaws must be turned, or bored, while applying the same pressures which will be applied in machining the actual job. Does that help?

Soft jaws are also helpful when chucking pressures need to be moderated. If you are machining thin walled parts the pressures required to keep the part from turning in the chuck may be enough to distort the part. If this is the case then making, or purchasing, a set of "pie" jaws may be the answer. Photo 7 is a set of these jaws which I have used several times on a repeat job.

Here is a good place for a reminder. When you machine a set of chuck jaws and, after use, remove them from the chuck, they lose their concentricity. Always re-machine them when you use them a second time. The top jaws on my 3-jaw, and probably on yours, are numbered so that they may be

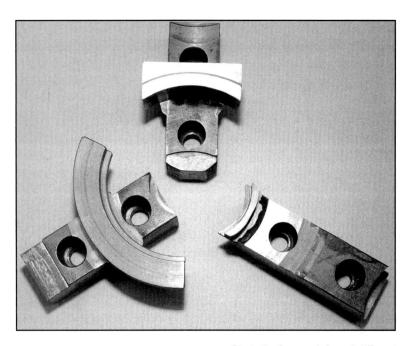

Photo 2. One each from 3 different sets of chuck jaws. Each set will of course have three jaws of the same configuration.

Photo 3. A stack of different diameter parts used to clamp in the chuck to machine the chuck jaws.

Building Shop   181

Soft Jaws

Photo 4. The setup for machining soft jaws. The sample plug is clamped in the chuck behind where the part will be held.

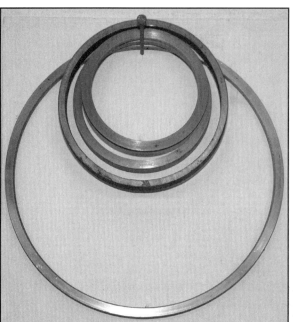

Photo 5. A sampling of rings which are used when the chucking pressures are from the ID.

put back in the same order in which they were originally installed. Numbering your soft jaws will do no good if you really want to take advantage of the system. It won't hurt but you should always re-machine them after they have been removed and replaced.

A discussion about soft jaws is not complete without mentioning pot chucks. A pot chuck is merely a device for chucking small parts and is used with a collet system. The ones shown here in Photos 8 and 9 are based on the 5-C collet system. Photo 8 shows two sizes of pot chucks, both of them machined to fit the 5-C drawbar on my lathe. In Photo 9 you see one of them installed in the lathe. Notice the plug in the center of the collet. This plug does the same thing as the sample part used to machine the diameter of a soft jaw. You tighten the collet on the plug, machine to the diameter of the part you are going to be working on, then remove the plug and chuck your part. It is a very effective, fast and accurate system when you are machining multiple parts. Expanding collets are also available if your requirements are for machining the OD of small parts.

This brief article about machining with soft jaws is like most such articles. It only touches briefly on a subject which has many facets. And also like most articles about machining processes, it ignores completely the endless ingenuity of the machinist and the craftsman. That is where the true diversity of our profession or hobby becomes manifest. And if this article about soft jaws doesn't have anything in it to interest you, then go call up your bird dog, shoulder your favorite shotgun and go bird hunting.

Photo 6. One of the rings in place preparatory to machining the OD of the chuck jaws.

Photo 7. A set of chuck jaws which provides almost 360 degrees of clamping pressure. Useful when the part is thin walled.

Soft Jaws

Photo 8. A pair of pot chucks. These are used with the 5-C collet system and are useful for repeat operations on small parts.

Photo 9. A pot chuck in place to be machined to accept the part. Notice the plug in the center which must be removed after the chuck is bored to size.

Chapter 16
Sweeping the Head of the Turret Mill

Sweeping the Head of the Turret Mill

There are many old jokes about sending the apprentice to the tool room to fetch the "left handed hammer" or to get a spool of "National Fine Threads," or a "sky hook" or some other fool's errand which in many shops were just a part of the daily routine. I have been the subject of a couple of them myself. One such joke is to hand the apprentice a broom and instruct him, or her, to sweep the head of the milling machine. I have taught many students and apprentices how to perform this simple, basic shop maneuver and have stressed its importance to them all but it is sometimes still ignored or over looked and the result will usually be puzzling over why the part didn't fit or why did the holes not line up. The process is probably easier to do than to explain but I am going to try here to answer the question I know you discuss daily at the dinner table with your family and wonder about before you go to sleep at night. Why must you "sweep" or "tram" the head of your vertical milling machine and how do you do it?

Why do it?

In most of the operations we perform on the vertical milling machine the spindle must be exactly perpendicular to the table. Photos 1 and 2 help to illustrate this. I have adjusted the position of the head of my machine for purposes of illustration so that the top of the head leans to the left when viewed from the front of the machine. In

Photo 1. Taking a cut with the head out of adjustment side to side. Notice that the cutter is only cutting on one side.

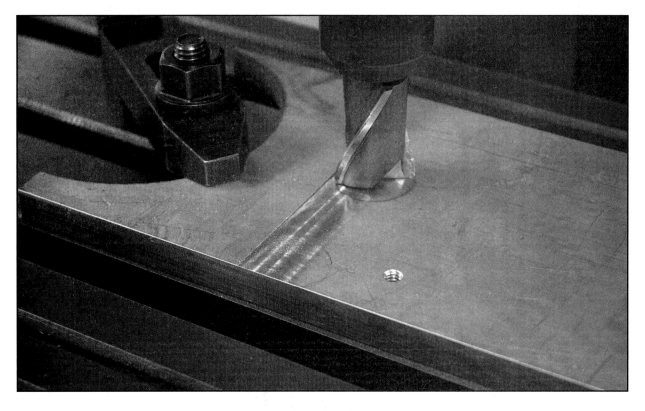

Photo 1 we see that the cutter is cutting on one side only. Photo 2 shows what happens if you try to take successive cuts across a plane surface. The result is a series of ridges sort of like a saw tooth arrangement. You can see that the more the spindle is off, in either plane, the more pronounced the effect will be. Another problem brought on by an out of perpendicular spindle is this: If you indicate a part or a feature of a part at a given elevation of the knee, and then raise or lower the knee to a different elevation, your spindle becomes misaligned from its reference point. The way we avoid these problems is to "sweep" or "tram" the head of the milling machine.

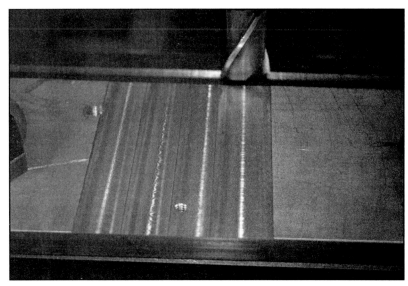

Photo 2. Taking several cuts across a plane surface. The cutter leaves a series of steps along the direction of the cut.

How is it done?

Begin by taking a good look at Photo 3. This is the head of my Seiki-XL milling machine but the arrangement will be similar to

Photo 3. A picture of the head of my Seiki-XL milling machine. The points needed for tramming the head are labeled.

Sweeping the Head of the Turret Mill

Photo 4. Using a large diameter plate for an indicating surface. The plate in the picture is a part sawed from an aluminum billet. If you use this method you must have a plate whose sides are machined parallel.

The likelihood of a tap breaking is directly proportional to the percent complete of the job you are doing. You will never break a tap on the first hole.

that of most small machines referred to as the Bridgeport type of milling machine. Points A and B indicate two scales and witness marks. Set both of these to zero for a starting place. We will be coming back to this photograph so don't lose your place.

Everyone develops preferences in how to go about this process. My own preference is to clean off the machine table and use it for a reference plane. Some use a plate, see Photo 4, some use a fabricated device which is said to allow tramming the head while the vise is mounted to the table and there are probably others. But if you go directly to a clean table you eliminate the possibility of there being burrs or chips between any of the surfaces. By the way, the plate in Photo 4 isn't suitable, anyway. If you use a plate it must be machined, preferably ground, parallel.

The head of this machine swivels in two directions; from side to side and from front to back. We will first adjust the side to side motion. Photo 5 is of a dial indicator mounted in a Jacobs chuck in the spindle of the machine. Tighten the quill clamp and take the spindle out of gear before you begin. Since the indicator button has to pass over the T-slots in the table I go about the process this way: Elevate the knee until you get a positive reading on the indicator. Set the indicator to zero and then set the collar on the knee feed crank also to zero. See Photo 6. After making both these settings, lower the knee until the indicator button is free, rotate the spindle 180 degrees and raise the knee back to zero. If the indicator doesn't read zero you are out of adjustment. Now go back to Photo 3. The four clamping bolts, B1, B2, B3 and B4 are on the front of the head and must be loosened to make this adjustment. Loosen the bolts and turn the adjusting

Photo 5. Mounting the indicator in the spindle. Tighten the quill clamp before beginning the adjustment process. It also helps if the spindle is taken out of gear.

screw D to move the head. If your indicator reading was .010", turn the screw until you read half of that, or .005". Zero the indicator again, lower the knee, rotate the spindle 180 degrees and take another reading. Take one final reading on each side after the clamping bolts have all been tightened to make sure and there you have it.

That was the easy one. The other adjustment, front to back, is a little different and must be done a little differently. Note in Photo 3 that the pivot point, point E, for this adjustment is located about 8-1/2" from the center line of the spindle. We will go about making this adjustment in essentially the same fashion but it is not so straight forward as the side to side adjustment where the indicator readings are equal from one side to the other. But the process and the desired end result is the same. Loosen the clamp bolts, A1, A2 and A3, adjust the head by using the adjusting screw, C, and proceed as before. You want for the indicator readings to be zero-zero on the front and the back of the table.

A few notes:

- The witness marks will allow you to get to a starting point but never depend upon them to get you exactly in adjustment.

Sweeping the Head of the Turret Mill

Photo 6. The elevating crank set to zero.

- Use as long a horizontal rod as you can when mounting the indicator. A longer rod will amplify the error making it easier to detect and adjust.

- When adjusting the front to back tilt of the head it is easier to adjust up. The weight of the head makes adjusting it down smoothly difficult.

- On some machines, particularly larger ones, a heel angle is required. Milling with large diameter face mills can leave undesirable marks on the work on the back side of the cut. Including a slight heel angle can eliminate this but it is probably not useful on this type of machine. Just so you know.

- If you are the sole user of your machine, frequent adjustments are probably not necessary. But check it once in a while anyway. You probably never have accidents with your milling cutters but if you do, the head can be knocked slightly out of alignment and you won't know it until you have messed up a part.

- And that is pretty much all there is to it. This is a pretty simple procedure but it is often overlooked and, I think, sometimes misunderstood.

So if your apprentice insists on using a broom for this process, leave him alone. My milling machines usually can use a little tidying up.

I love having visitors in the shop but call before you make the trip. If I am somewhere else then I am not here.

Chapter 17

Not Just a Lathe — A *Contouring* Engine Lathe

Not Just a Lathe

When I was working as a tooling machinist many years ago in an oil tool manufacturing facility I was introduced to the tracer lathe. They were pretty neat devices which enjoyed a short span, or maybe a long span, of popularity before CNC made them effectively obsolete. They consisted of a template mounted on the back side of the machine with a two-dimensional profile of the part and a stylus which followed the template, producing a tool path which duplicated the profile of the template. It worked a little like a taper attachment but with an additional degree of freedom if that is a valid way of describing the process. They were pretty useful for their purpose and would no doubt still be in use today had they not been superseded by the computer.

The tracing attachment is what provided the inspiration for the attachment I am going to describe here. In Chapter 12, Building An Engraver's Vise, we talked about methods for machining spherical profiles on the engine lathe. This method is only one of many for accomplishing that purpose but it has the added advantage of not being limited only to spherical profiles. Any template you can saw out and mount on a spar on your lathe can be reproduced accurately using this method. It has its limitations, I will agree. But in the absence of more expensive alternatives it will do the job. Photo 1 and Photo 2 show the device being used on my lathe to machine the profile for the engraver's vise.

Photo 1. Roughing out the ball for the vise.

192   Building Shop

Begin by deciding how to mount a rigid spar on the back side of your lathe bed. It would be pointless for me to provide drawings or other specifications for the installation of the spar because there are so many variations of machines but here are the requirements:

Photo 2. Taking the finish cut.

It must be heavy enough to provide a rigid support for the template you will be mounting on it. On my machine I used a piece of hot rolled steel 1" X 2" in cross section.

It must be of sufficient length to allow the carriage wings to travel all the way along the profile. Remember that the carriage wings may extend as far as 8 or 9 inches on either side of the center of your profile. In Photo 3 you can see that I have made the spar long enough to allow for other templates besides the one I am using for this job.

It should be easily attached and removed from your lathe. The easier it is to use the more useful it will be to you. Photo 4 shows the mounting posts I made for this application. They are readily installed and readily removed from the lathe.

Photo 3. The spar and template installed on the lathe.

Next make the follower arm and here again you will have to make some choices. I made the follower arm you see in Photographs 5 and 6 to mount on the side of the compound. It worked well here for this job and it was easy to remove the compound and drill and tap the holes on the milling machine.

Building Shop 193

Not Just a Lathe

Photo 4. Right, mounting bracket for the spar. There is one at each end.

Photo 5. Below, cam roller and follower arm. Shown upside in this photo.

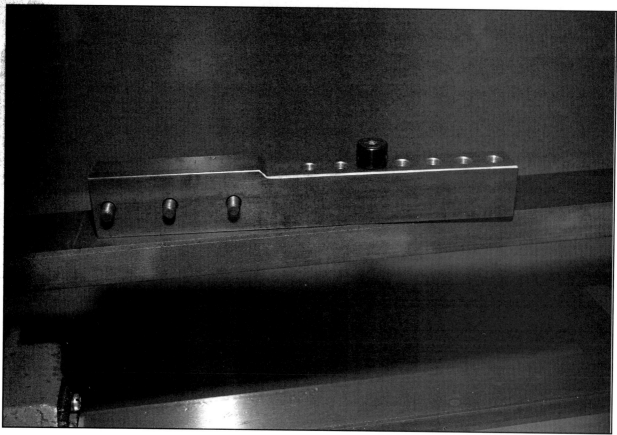

Photo 6. Cam follower in use.

There will be many times when a follower mounted directly on the cross slide will work better. Here again, you must evaluate the choices available to you depending upon your equipment and methods. In Photo 5 you can see the cam follower bearing I mounted on the bottom of the follower arm (shown upside down in the picture) and you see also several mounting holes I drilled and tapped to allow for a greater range of diameters I can work with.

In Photos 7 and 8 you will see the template made for this job. It is machined from ¼" steel plate but a lighter material would probably work as well. It must be rigid enough not to bend under the pressures applied by the cam roller.

And that, essentially, is all there is to it! After making and installing the spar and the template here are some suggestions for their use:

Chuck the work piece in the lathe and move the tool to touch the end of the part. Then move the tool so that you can touch the outside diameter at the headstock end of the profile. Make sure that you have sufficient clearance to move the tool all the way through the desired profile without having to reset the tool midway through the job.

Set the compound at 45 degrees. It must be at an angle to both axes because if it is parallel to either lathe axis, feeding it in that axis will not result in any material removal.

Just what, exactly, do people mean when they tell you they have a "full set of twist drills?" That would be an awful lot of drills!

Photo 7. Template mounted on the spar.

Rough out the profile, leaving a comfortable amount of material for the finish cut. You can do this by retracting the compound by the amount you want to leave for a finish cut and then advance it that final amount when you are finished with the roughing process.

I find it works best on most parts to hand feed the machine in both axes, particularly while roughing. For finishing this part it worked well to engage the X-axis feed feeding out from the center and maintaining a pressure against the cam follower and the profile by hand. Working the other way may work better for you but once more, it depends upon your material, your cutting tool, your machine and your state of mind.

As we have said before, this is just one method for producing profiles other than square and linear on your lathe. It is a handy thing to know and it can add to the capabilities of your shop without adding a lot to your expenses. It will also give you some bragging rights in some company but your wife may not be impressed.

Photo 8. Another view of the template.

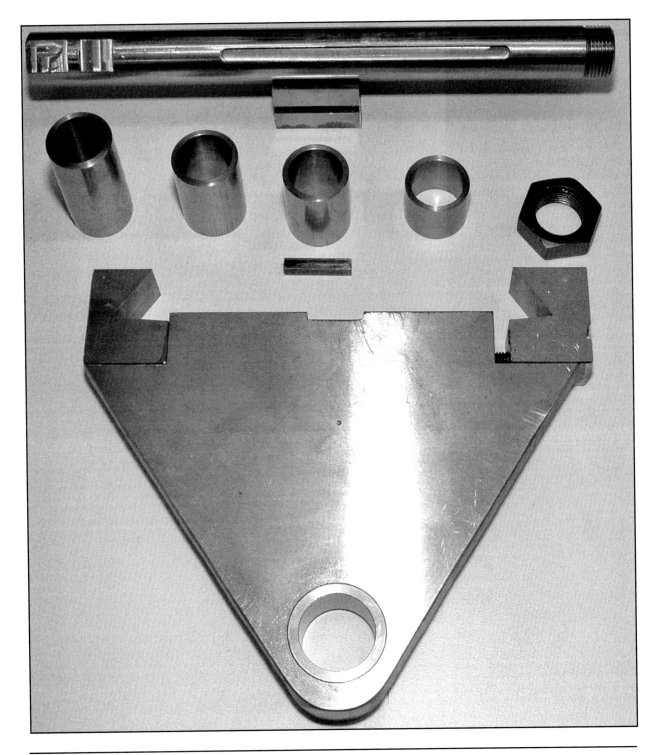

Chapter 18
A Horizontal Milling Machine — On End

Horizontal Milling Machine On End

As promised back in Chapter 5 here is a description with some photographs about converting your vertical milling machine to a horizontal milling machine — on end. This article was completed and submitted to *Village Press* in June of 2008. Since that time I have had more occasions to use the device than I thought I might. Maybe I am just looking for excuses.

The turret-ram vertical milling machine found in most of our shops is remarkably versatile, particularly when outfitted with one or more of the many accessories available for use with it. But some cutting tools, slitting saws and form cutters including gear cutters just to mention a few of them, require some sort of an arbor mounting system. Options are available. One option is the right angle milling attachment shown in Photo 1. If you are diligent, and lucky, you may come across one of these on eBay but don't count on it. This system requires a ninety degree gear head which mounts on the quill and an arbor support, or yoke, to support the opposite end of the arbor. A good system and certainly one to consider if it is available when you need it. Another option is to use one of the stub arbors shown in Photo 2. Both of these options are good and have their applications. But as is always the case there are other alternatives. The alternative I am going to describe here is a case in point. It won't work every time you need a supported arbor but it does work and it isn't difficult to make. How you go about it will doubtless depend upon the condition and

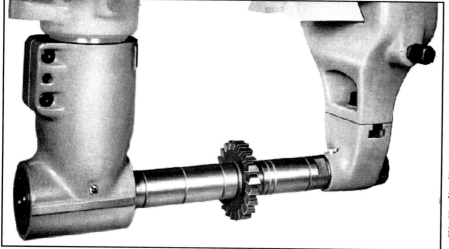

Photo 1. A Bridgeport horizontal milling attachment. This consists of a ninety degree head and an arbor support, or yoke, which clamps to the dovetail of the ram.

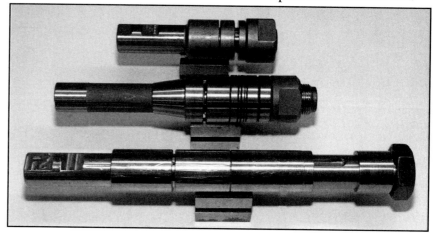

Photo 2. Right, three cutter arbors. Note that the one in the middle is for an R-8 spindle. The others mount in a Weldon tool holder.

contents of your scrap pile (universally known as the alternative material storage facility) and the tools you have available. Whatever system you choose or devise to mount and drive an arbor mounted cutter on a 2 HP milling machine isn't going to allow you to run a 6" slab cutter but if we aren't able to accept some limitations we have no business being in the hobby we are in. Right?

Photo 3. The completed system in use cutting a gear.

Photo 3 shows this system in operation. I am machining here a 40 tooth, 20 DP, 14-1/2 degree spur gear. This is a good setup. The tool arbor is supported at both ends and the work piece is also well supported by using the foot stock with the indexing head. Things get a little close together here down at the bottom but if you have adequate clearances between all components of the setup then it really is true that an inch is as good as a mile. If you are interested in applying this method to your machine here is how I did it. I am not including drawings because there are too many differences between machines and applications. I hope the photographs will convey the idea.

Photograph 4 shows all of the component parts required to make this work. I used some of the keyed spacers from a stub arbor to complete the arbor assembly after I had chosen a cutter. Begin by measuring the angle of the dovetail on the column of your machine as seen in Photo 5. My machine measured 35 degrees but I don't know if that is a universal angle for these machines. Lay out the profile on a piece of 1" thick plate and band saw as shown in Photo 6. I used a piece of low carbon steel plate which had been lying around. I see no reason why a piece of mild steel or even 6061 aluminum plate would not work. The 1" thickness is also arbitrary. I would not recommend using

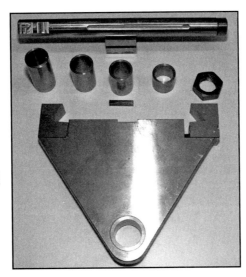

Photo 4. All components required for converting a vertical mill to an arbor supporting system.

Photo 5. Measuring the angle of the dovetail of the column.

anything lighter than about ½" because of the mounting method but a weldment would certainly not be a bad choice. Using a ¼" steel web with bosses welded in where the device mounts to the column and where the arbor bearing goes would be a good option if you so choose.

To achieve the required 35 degrees I used a 30-degree cutter and set the head of the milling machine over 5 degrees as may be seen in Photo 7. It would have made a better job to have been able to cut the 35 degree angle and leave the left side of the plate in one piece. But I don't think the resulting loss of bearing surface will affect the finished job. Alternatives would be to use a shaper or a vertical slotter to machine the dovetail or even a carefully band sawed part might work. I don't think you would want to flame cut the parts but I don't know how good you are with a cutting torch so I make no recommendation there.

After completing the machining of all of the column clamping surfaces, set the parts up in either the milling machine or a drill press and drill and tap the holes. See Photo 8. Then clamp the plate to the column of the machine. Make sure it is square and level in relation

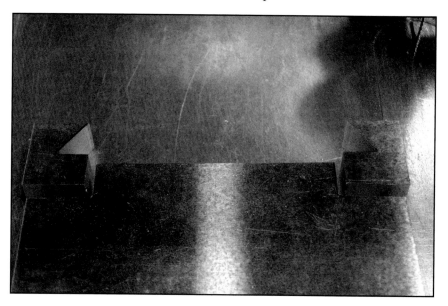

Photo 6. The arbor support plate sawed from 1" plate.

to the machine and then move the ram to the position you will be using when the setup is complete. Refer to Photographs 9 & 10. I set the turret at zero degrees and the ram at zero inches. Both are easy to remember for the next time I use the yoke.

Begin by drilling through the part with a large diameter drill. Notice in Photographs 11 and 12 that I have placed an angle plate under the outer end of the plate. Some support

Photo 7. Cutting a 35 degree dovetail with a 30 degree cutter.

is a good idea here particularly during the drilling process. Put the support in place and then elevate the table until the plates are in firm contact. After drilling, bore to the desired diameter to accept a bronze bushing. Use either a finished bushing or press in a piece of bronze and bore it to its finish size. Keep the support in place until you have completed the boring operation. Here is something

Photo 8. Drilling, tapping and counter boring the attachment bolts.

to keep in mind. It is handy for the inside diameter of the bushing to be larger than any of the spacer bushings used on the length of the arbor. That makes setting up a little easier. But remember that things get close down there among all those setup features. The arbor support plate, the milling machine table, the nut on the end of the arbor, the foot stock for the index head (if you are using one) and all the other associated parts required for setting up the job. The smaller you can make the plate out at the bushing end, the more convenient it is for setting up. A shielded ball bearing here would be a good thing but the bronze bushing will probably serve for as long as I need it and the bronze bushing will take up less space where space is already at a minimum. This is another of those compromises we have to make from time to time. Don't make it too small. Just make it fit. The last thing to do on the arbor support plate, or yoke if you want to call it by its correct name, is to trim away all of the unnecessary material. Keep in mind what we were just talking about. Make it as small as you can without compromising the strength.

The carpenter's adage of "Measure twice, cut once," isn't always a good rule. I will sometimes measure five or six times before I am ready to cut. Depends on what you paid for the material.

Now for the arbor itself. If you are a glutton for self punishment, you can machine an arbor with an R-8 configuration built in. That is OK but you will limit yourself to whatever length and diameter you settle on. Whereas, if you use a straight arbor and a Weldon tool holder, making new arbors to fit the next job will be much less time consuming. The middle arbor in Photo 2 is an R-8 arbor, 7/8" in diameter and I don't even remember why I bought it. I do remember it was for a particular job and it served its purpose but the straight arbors with the setscrew flats machined on them serve equally as well. To make an arbor, first determine the bore of the cutter you will be using and select a piece of material to suit. I used 1" drill rod for the arbor in this example and that works well but any good piece of TGP (turned, ground and polished) steel will serve as well. Providing it isn't so hard you can't machine the flats, keyway and threads on it, of course. Machine the spacer bushings to random lengths and remember that the bushings on either side of the cutter and the bushing which will turn in the bronze spacer should have keyways in them. And that is pretty much all there is to it. Your own application should dictate changes to what we have talked about here. You may, for example, want to use a larger plate and move the bushing further from the machine column. There may also be applications where you could need the lower arbor mounting point to be offset to one side of the knee or the other. You can put it wherever you want for it to be but remember that setting the machine up for subsequent operations will require dialing in the bushing. Which is not a big job at all. Just set the turret and the ram both back at the settings noted when you made the arbor

Horizontal Milling Machine On End

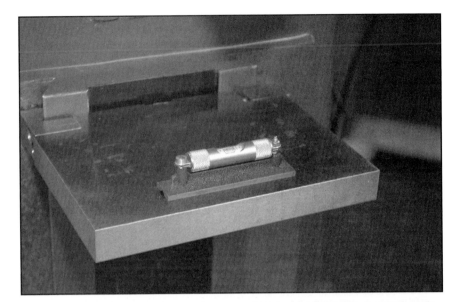

Photo 9. Leveling the arbor support plate preparatory to machining the hole for the bronze bushing.

Photo 10. Settings I have chosen for locating the turret and the ram.

support and then use a dial indicator as shown in Photo 13.

This project is a perfect example of what I preach to every congregation who will listen to me. (Which aren't many, by the way!). The only limit to what can be accomplished in the machine shop is placed there by the machinist himself. Machine capacities, tooling availability and materials all will influence the outcome but an imaginative machinist has no limits to what he can do given the time and resources. If it is true that necessity is the mother of invention then I submit that the capable craftsman is the father. But don't ever think you are breaking new ground. I will bet that I hear from at least a dozen people who have used the process I have described here and over half of them will have ideas about how to improve on it. We are never as smart as we think we are.

Horizontal Milling Machine On End

Photo 11. Drilling the hole. Notice the support plate under the outer end of the plate.

Photo 12. Boring to finish size for the bushing.

Photo 13. Process for re-installing the support plate after it has been removed.

Chapter 19

The Wonderful World of Watts
(Watts Tooling, That Is)

The Wonderful World of Watts

It is an inescapable fact of life for a machinist that we will always discover something new concerning our work at about the time we think we have seen it all. Watts tooling is a good case in point. I discovered Watts tools back in the early seventies but I remember thinking at the time, "How could I not have known about this stuff?" It always makes me wonder what other basic sort of tool or method there is out there that I have not yet been exposed to. And believe me, there is more out there.

This system was the subject of an article which appeared in the April/May, 2008 issue of *Machinist's Workshop* and is a system I find frequent uses for in my shop. It makes one wonder what we did before we knew about some of these things. Did we design or implement alternatives? Or did we just do without?

Watts Brothers Tool Works of Wilmerding, PA was founded in 1916 and have been providing tools for drilling polygonal holes to the industry ever since. It is not the intent of this discussion to educate the machinist in the use of these tools but merely to make you aware of their existence and to make you wonder as I do, "Now that I know everything, what else do I not know?" If you do a Google search (Isn't it remarkable how the practice of "Googling" has so quickly become a main stay of our society?) look for Watts Brothers Tool Works. They do not have a prominent presence on the Internet that I could find but they are listed.

These tools may be used in a rotating spindle, as in a drill press or milling machine, or in a lathe where the work rotates. They work equally well in either application. The accompanying photographs provide a much clearer explanation than my words can. Do not try to buy just the cutting bits and use them in a conventional Jacobs chuck. It won't work! The minimum you will need for this system is the full-floating chuck and the guide plate and drill for the appropriate size and configuration of the hole you desire to machine. You will be required to make an adaptor plate or plates for the work pieces you are machining but if that is a problem for you then you probably should not be considering the tools at all.

The instruction booklet which comes with this tooling is clear and comprehensive. And, by the way, written in English! (Unlike so many instruction manuals we see today.) Photo 1 is a picture of the purchased tools required to machine either a square or hexagonal hole. Notice that the cutting tool for the square hole, shown on the left, is a three-fluted tool and that the tool for machining the hex is a five-fluted tool.

There is no excuse — none — for allowing a drill to penetrate through the work and into the vise or drill press table. I don't even want to hear it!

The Wonderful World of Watts

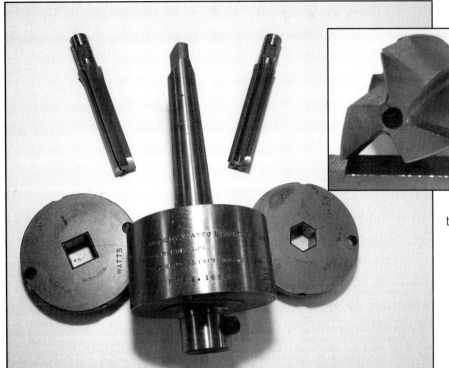

Photo 1. Purchased tooling required to machine both square and hexagonal holes. Inset, view of drill flutes

Photo 2 shows the adaptor plate I have made positioned on the end of the shaft. The adaptor plate is held in place on the end of the shaft by a setscrew. If the job requires machining holes on a flat surface using a drill press or a milling machine the adaptor plate may be attached by bolting it directly to the job. Photo 3 shows the next step in the process.

The Watts guide plate has been attached to the adaptor plate with two socket head cap screws making the part now ready to drill. A pilot hole is not required in soft material but I usually will drill one anyhow. The instruction manual will give recommendations as to pilot hole size and cutting speeds. Maximum depth will depend upon the size of the hole but a depth of about two times the distance across flats is a pretty good rule to follow.

Photo 2. User provided adaptor plate in place.

In Photo 4 you see the cutting tool in position to enter the work piece and Photo 5 shows the finished square hole.

As I noted at the beginning this is not intended to serve as a how-to lesson for Watts tooling. It is intended only to make the reader aware of the process. It is a simple and effective tool for machining polygonal holes and will help to give your projects a more professional appearance. Whether or not you reveal the secrets of how you do it is up to you.

Photo 3. Guide plate attached.

Photo 4. Ready to drill.

Photo 5. ½" square hole drilled.

As Will Rogers almost said, I never met a machine tool I didn't like. Well — there was that old 30" American lathe in that oil tool shop in California that time.

Chapter 20

From the Chip Pan

From the Chip Pan

Usually the last thing we do in a machining job is cleaning out the chip pan. Here you will find inserts, screws, Allen wrenches, motor blocks, back-hoe buckets and other items most of which will cause you to ask, "Now how in the pluperfect H--- did that get in there?" The chip pan is also a place where some items go never to be seen again. There is no explanation for it, scientific or otherwise. They are just gone. Like odd socks from a dryer. It went in but it never came out. The mystery will never be solved so we just accept this as a fact of life in the machine shop.

This chapter is like a chip pan. This is stuff that doesn't fit anywhere else and nobody knows much about where it came from. Just swarf and scraps which escaped being swept out when the cleanup crew came through on their regular pass through the shop. The first item here is a good example:

I once saw advertised a "Complete Assortment of Hardware." They said it had all the fasteners, pins, clips and hardware necessary for every maintenance need. I didn't order it because I knew I didn't have the 2,500 square feet of floor space it would require.

Making a Tail Stock from a Tail Stock

If you are going to be making a tail stock for a lathe then having a tail stock from which to start is probably a pretty good thing. We were building a wood lathe for my brother by converting an old second operation Rivett turret lathe. I machined a new spindle so that the little machine could be used for turning between centers as well as providing the ability to turn larger diameters out on the left end of the head stock. Making the tool supports was a pretty easy job and we mounted a new motor and jack shaft to provide a good range of speeds for turning wood. The head stock bearings were more than adequate and this was going to turn out to be a pretty nifty little lathe for wood turning. Except it had no tail stock!

Photo A1. The completed tail stock mounted on the bed ways.

The turret, which was originally a part of the little machine, had already found a new home and besides, it took up too much of the bed length to be considered as a substitute for holding the tail center. So we had to make one. Photo A1 shows what we eventually came up with. Fabricating one from scratch was considered and that was still an option, but a little searching on eBay turned up

Photo A2. A rear view of the tail stock. Note the key machined integral with the base of the unit. This will obviously depend upon the configuration of the machine.

a good alternative. We found an orphaned tail stock in pretty fair condition from an old 11" lathe. Its pedigree was uncertain but it was also unimportant which made shopping for it a little simpler. It was old enough to have been cast with the white lead reservoir in the top and most importantly, it was complete. The hand wheel and the spindle locking handle were present and the threads in the barrel were worn as might be expected but not badly at all. So now we begin the job at a point where it was already over half completed!

The first step was to make the adapter plate which provided the transition from the configuration of the top of the bed of the Rivett to what would be the bottom of the tail stock. This was made from a flat piece of steel of an appropriate size. See Photos A2 and A3. After determining what the height of the base plate is, now measure the existing height of the tail stock in relationship to

Photo A3. Machining the new base plate for the tail stock. The original plate was discarded. If height is an issue you may need to keep the old plate and machine a new bottom for it. Or you may even need to add more height to the job. Depends upon many variables.

From the Chip Pan

Photo A4. Measuring the height of the centers. What you require here is the difference in the heights of the two points. This gives you the information for fabricating the base plate.

Photo A5. Marking the casting for sawing. Stack parallels to give you the line for sawing.

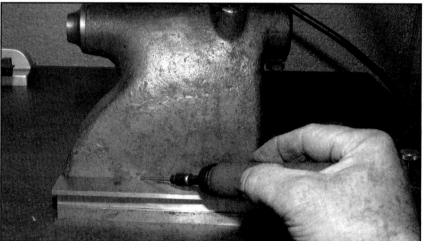

Photo A6. Sawing the base from the casting. I have bolted the unit to an angle plate. Note the use of blocks supporting the casting. This makes the job less likely to tilt in the saw and possibly break your blade.

the centerline of the spindle. Photo A4 shows one method of doing this. If this were going to be the center for a precision application closer measurements would be required but for a manual wood lathe this method is close enough.

Now place the unit on a level surface and scribe a line where the bottom of the casting will be cut off. You will see in Photo A5 that I have stacked parallels up to use as a straight edge for marking the part. Leave enough material when you saw to allow you to finish the surface in the milling machine. I bolted the part to a ninety degree angle plate to hold it for sawing. The sawing process may be seen in Photos A6 and A7.

Here is where you will have to be a little innovative. Notice in Photo A8 how I have set the part up in the milling machine. I used a piece of cold rolled shafting the same diameter as the bore

212   Building Shop

Photo A7. A good example of why you need a large capacity band saw in your shop.

of the tail stock quill and clamped it in a pair of V-blocks to the milling machine table. Then use jacks and clamps to secure the part tightly enough to resist the forces you will apply when machining the bottom. This worked very well and if you do require a higher level of precision for the tail stock you can achieve that here.

Now all that remains to be done is drill and tap for the bolts which attach the tail stock to the base plate, see Photo A9, and machine a clamping nut or plate nut which you can see by going back to Photo A2.

Photo A8. Machining the bottom of the casting. If you require a precision height for your tail stock this setup will allow that. The bar in the V-blocks insures the bore of the tail stock will be parallel with the base.

Photo A9. Drilling and tapping the holes for attaching the base plate. Use enough bolts to insure the assembly but when the tail stock is clamped to the lathe bed there is virtually no strain on the bolts.

Making a tail stock from a tail stock turned out to be much simpler and less expensive than making one from the scrap bin and, in this case anyway, made for a good job. That is always a worthy goal.

From the Chip Pan

Making a Tool Block for a Quick Change Tool Post

Here is a project which is frequently discussed on the bulletin boards and in other places. It may even be discussed at the Congressional level but for the fact that this is actually a project which can be completed on time and under projected costs. I don't think that ever happens in Congress. But maybe I should stay away from those comparisons.

The wedge or piston actuated quick change tool post has become popular for use on manual lathes and it also will be found on many CNC machines. It provides a quick, easy and accurate method for changing from one tool to another and back again with a high level of repeatability. There are several makes available and it is convenient that many of them can use the tool holders interchangeably. This is not to say there aren't other and better options but I learned years ago not to go that far out on a limb. I will just say these tool posts are what I use on both my lathes and they serve my purpose — except when I don't have the proper tool holder to go with it. That is what I am going to talk about here.

I have an A series Aloris tool post on my Hardinge lathe, a B series Armstrong tool post on my Kingston lathe and a C series Dorian tool post stored in the back room. The Dorian came to me by way of an auction and will probably leave the same way as I do not have nor need a lathe big enough for that size tool post. But, you never know!

The problem was caused by the tooling I use for parting. I use the Self-Grip system devised and sold by Iscar tools. It does what I need for it to do. The problem is that the Iscar blade holder is 1" high and neither my Aloris nor my Armstrong tool holders will accept a tool that big. So I had to make some.

Photo B1a and B1b. Two tool blocks made to accept Iscar parting tool holders. One for the A size tool post and one for the B size.

Photo B1 is a picture taken of the two tool holders I made, one for each size of tool post I use. They both will accept the Iscar 1" blade holder. They are made from 4140 alloy steel and heat treated to 40-42 Rockwell C hardness and they have paid for the cost of making them many times over. For a while I got by with only having one for the small tool post and if I needed a parting tool on the larger lathe I had to swap out the entire tool post. Now that I have one for each lathe life is a little easier. You always want to make life a little easier when you have the opportunity.

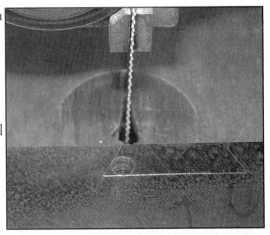

Photo B2. Preparing to saw the dovetail. Lay the lines out to leave about .03" for machining. Note the hole drilled to facilitate turning the corner.

Photo B3. The dovetail sawed out. This saves a lot of machining time.

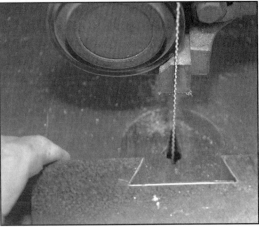

The photos included here are some I made while making the larger tool holder, the one for the B series Armstrong. Photos B2 and B3 show the part being sawed out of the blank. Sawing the dovetail out speeds up the process but be sure to leave a little material for finishing. I try to leave about .030" all around. Photos B4 and B5 are of the milling process and are pretty straight forward. But look closely at Photo B6. This shows a really

Photo B4. Machining the slot for the cutting tool. It makes no difference whether you machine this side or the dovetail side of the block first.

Photo B5. Machining the dovetail.

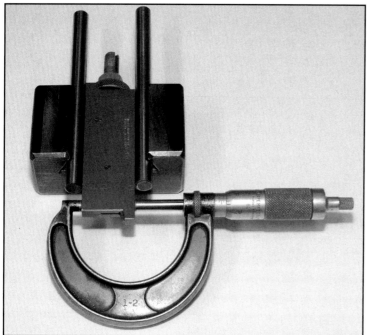

Photo B6. The preferred method for measuring the inside of a dovetail. When measuring the outside profile just use the hardened dowel pins. A little trigonometry will help you out here.

good method for measuring dovetails. Select a couple of pieces of drill rod, you can use drill bits for this or almost any appropriately sized dowels, which will fit into the dovetail and contact both the bottom surface and the angled surface. The ones shown here are 3/8" in diameter and were just right. Then open an adjustable parallel so that it is snug in the space between the dowels and measure with a micrometer the width of the parallel. You can go to your Machinery's Handbook and find the formula for calculating the exact distance between the sides of the dovetail but you don't really need to know that here. All you need to know is that you want both the pattern and the one you are making to be the same. If you do not have a pattern you will have to do a little head scratching trigonometry.

After machining the tool slot and the dovetail all that is left is drilling and tapping for the setscrews and for the height adjusting nut. You can go as far as you like in making this tool holder exactly like the one you used as a pattern. I machined a knurled nut and made my 1" tool holder exactly like the ¾" tool holders I already had for both sizes of tool post. You might want to do the same.

The time and effort which goes into marking your work is almost always worthwhile. There are many ways of doing this and the method I show here is one which will result in a nicely professional appearance. Photo B7 shows a neat but shallow groove put into the work piece with an end mill. I have a stamp with which I mark my

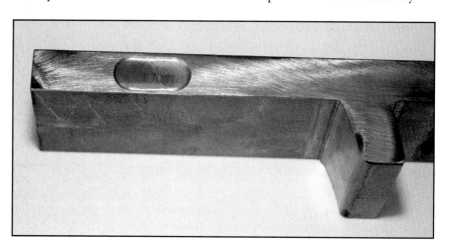

Photo B7. An identification pocket machined in a part.

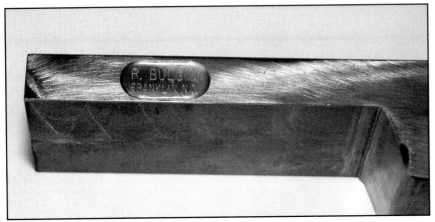

Photo B8. An example of marking your work. A good practice to develop.

Photo B9. An alternative method for marking if you do not have a custom-made stamp.

work, mostly ornamental iron work I have done, and I will usually use it for machined parts. (I don't stamp parts which are out of tolerance or otherwise not suitable. I only mark the good stuff!) Photo B8 shows the part marked with my stamp. The V-block in Photo B9 is one I made many years ago before I had my stamp and I used the same process but with individual letter stamps. Either way it looks nice and is worth the effort.

The last thing to do is to heat treat the part. If you have made this from a piece of steel which is already heat treated, and that is an acceptable alternative, you won't have to do this. But I prefer tool blocks to be a little harder than what is available pre-hardened. Pre-hardened steel will be at approximately 28 to 32 Rockwell C. I tempered these blocks to 42 Rockwell C. Heat the parts to 1575 degrees F and quench in oil. Then temper at about 650 degrees F. Wrapping the parts in stainless steel foil will help to reduce the oxidation and I use it for nearly everything I temper. Be sure that if you are going to mark your work that you do it before you do the heat treating. Stamping doesn't work really on parts which are harder than the stamps.

I now have more tool blocks for my lathes which I have made than I do purchased ones. I didn't plan for it to be that way. I have just been at this a long time and things accumulate. I wish I could accumulate cash as readily as I do tool holders!

Photo C1. Six screwdrivers I have made including a couple which haven't been heat treated.

Making Screw Drivers

Why on Earth would anybody spend time making screw drivers when they are available almost everywhere and for peanuts?! Simple. Because I can make a better screw driver than I can buy and I like owning good tools. I am not going to spend a lot of time making a screw driver which will be used to open a can of paint or to change a wheelbarrow tire or to punch a hole in the bottom of a barrel but if I am going to be working on a quality gun or some other device which I don't want marred up with screw driver tracks on it I prefer a screw driver I have made. And I know I can buy quality screw drivers from several sources. I still like mine better. At least the ones of the sizes I am talking about here.

Photo C2. The raw material. Several pieces of ¼" hex stock of 4140 alloy steel. You can machine your own hex material but this saves a lot of time.

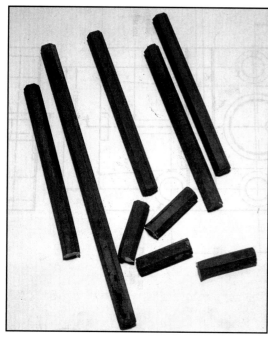

I made all of the screw drivers described in these photos and paragraphs from ¼" Allen wrenches. I have made a few bigger than this but there aren't many jobs which require specially fitted screw driver bits any bigger than these are. Photo C1 is a picture of several screw driver bits I have made, some of them quite a long time ago. Photo C2 is the raw material. Several pieces of ¼" Allen wrenches which have been annealed in preparation for use as these bits.

Photo C3 shows just two of the tools available for driving the bits. A ¼" ratchet and a ¼" socket driver. There are many ways of doing this and a good one is a magnetic ¼" socket. The disadvantage of using the magnetic tool is that you run the risk of magnetizing your tiny screws and that is not a good thing.

The first thing you will have to do if you are using Allen wrenches is to anneal them. This is done by heating to around 1550 degrees F and then cooling them very slowly. What works well for me is to wrap them in stainless steel foil (keeps down oxidation) before putting them in the furnace and then when the required heat is reached I turn off the furnace and leave the parts in it over night.

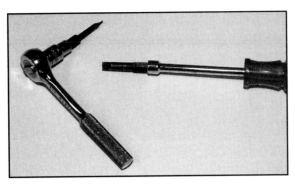

Photo C3. Two methods for driving the screwdriver bits. There are others.

After annealing the material, turn the ends in the lathe to the required diameter. This will equal, of course, the width of the screw driver blade you want when you are finished. In Photo C4 notice that I am using a radius tool to machine the diameter of the bit. Sharp corners should be avoided in this area. Photos C5 and C6 show the process of milling to thickness and Photo C7 is a picture of a useful device I have for my milling machine. When machining tiny parts like the ones we are working on here there is a tendency for the part to deflect. The fixture in the picture is a tool rest I made one time for use on a wood lathe. But it will never see a wood lathe again because it is so useful in this application. It is adjustable as you can see and by placing it to bear against the back side of the part being machined deflection can be almost entirely eliminated.

Photo C4. Machining the diameter of the bit. Using a radiused carbide turning tool for this job. Sharp corners here will weaken the screwdriver bit.

After you have finished with machining the screw driver bits, or whatever else you might think of making using this process (there are many possibilities) you will have to re-temper the tools. Wrap them again in stainless steel foil and temper them according to the process discussed for tempering the tool blocks.

Photo C5. Left, machining the bit thickness. It is important to keep the blade of the bit centered by machining equal amounts from either side. Note the use of a lateral support for minimizing deflection.

Building Shop 219

From the Chip Pan

Photo C6. Right, the bit almost finished.

Photo C7. Below, a picture of the support for machining slender cross sections in the milling machine. This is useful for many applications.

220 Building Shop

Repairing Cast Iron

The last item found here in the chip pan is a subject so diverse and so universal I almost hesitate to bring it up. Cast iron repairs will be with us as long as cast iron exists and that will be pretty much forever. Or at least until long after I am gone. Opinions about repairing cast iron; can it be repaired, can it be welded, will repairs to cast iron last, should we even try, etc.; will also be around for a long time and will never be universally agreed upon.

I am not going to burden you here with any of my opinions. These examples are jobs which have come into my shop for repair and these,

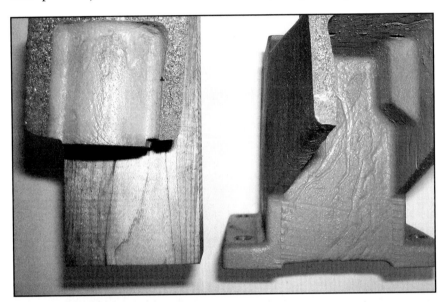

Photo D1. The first of three examples. This is the part as it came to me for repairs. The heaviest part of the broken cross section is about ½" thick.

Photo D2. The first step of the repair. Drilling for the insertion of studs which will be welded into the repaired area. These studs will add strength to the weld.

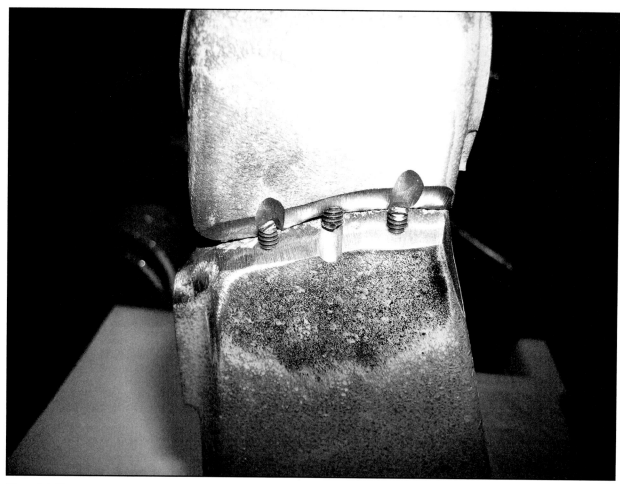

Photo D3. The studs in place. Note that we must grind clearance for the studs so that they do not interfere with fitting the broken parts together.

and many others like them, are a large part of what a job shop can expect. So don't argue with me about whether or not these jobs can be done because I have done them. The very next job to come my way may be the one I can't do anything with. I won't write about that one.

Cast iron has a bad reputation for welding because so many people don't go about it properly. It is just one of those things about human nature that causes most of us to blame our inability to master a process on the process itself. (If you knew how much time I have spent cussing Microsoft!) But there are many examples of cast iron repairs successfully made and will be many more. I won't try to make you believe that every cast iron job I have undertaken was successful. But I have a pretty fair record and a part of it is attributable to being selective in taking on jobs. Here are a few rules I try to follow:

- Stay away from exhaust manifolds from cars with over 200,000 miles on them.
- Don't try to weld motor blocks that are still in the car and which still have coolant in them.

Photo D4. The job clamped securely into position for welding. It is always a good rule to weld a little on each side in turn to minimize the distortion. Welding temperatures *will* distort the job.

- Suggest your customers try to find replacements for their stove grates.

- Don't give money back guarantees on cast iron welding jobs.

- Don't believe everything the electrode salesman tells you. Especially when he just tells you but won't show you.

- Don't take the job just because the customer brags on you. This applies to every job and not just cast iron welding jobs.

The accompanying pictures are of some jobs I have taken and had some success with. As I have already noted, I am not including photos of the jobs which didn't turn out so well. Many cast iron welding jobs will include some subsequent machining and that isn't a bad thing. After all, we are talking mostly about machine shops here, aren't we?

From the Chip Pan

Photo D5. Above, the finished job. A little grinding and some paint and you would never know it had been broken.

Photo D6. Right, the second example. This was from the same customer but I don't think the two parts were a result of the same incident.

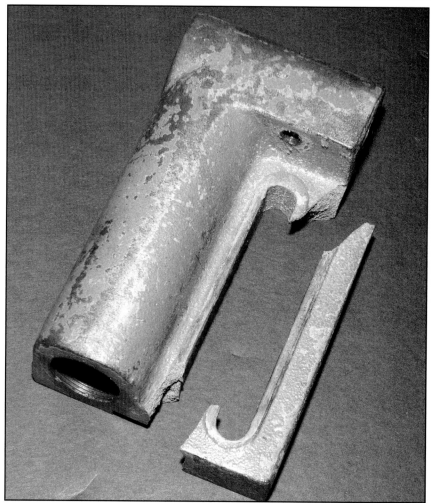

From the Chip Pan

Photo D7. A little different approach on this job. We will drill and tap for bolting the parts together. This does a couple of things. It holds the job in alignment while it is being welded and the bolts will add strength to the repaired joint.

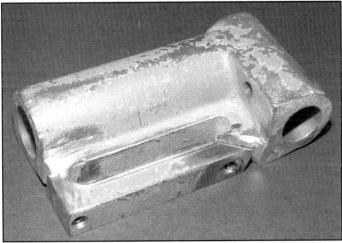

Photo D8. Ready to weld. Grind the weld prep joints after installing the bolts. Be sure to leave enough of the original broken surface to line things up before welding.

Photo D9. Repairs completed. The bolt heads are recessed so as not to interfere with any other part of the machine.

Building Shop

From the Chip Pan

Photo D10. And the last example. In this case the decision was made not to try to repair the part but to make a new one.

Photo D11. The replacement part. If you choose this option discuss it with your customer first. The grease fitting in this example was moved at the customer's request. The replacement was made from steel and will likely last longer than the original.

Photo E1. A stream of Teflon could present a danger.

And Finally...

Chip control. Who needs it? The photographs on the following pages are two examples of what we machinists sometimes have to put up with. Photo E1 is a picture taken while I was machining a large block of Teflon. Although the stream of material looks like it has gone across the crane beam it actually has not. The biggest danger presented by this stuff is that the Teflon has a fairly high strength even in the cross section of this stream of waste. It can get wound around the feed rod of the lathe or other rotating parts and really cause a problem. I guess if you were not paying attention it could even get wound around your neck but hopefully the level of awareness will be high enough to take steps to avoid that.

Photo E2 is of a scene not really that uncommon in my shop. I make a lot of parts from heat-treated 6061 Aluminum and it presents its own set of problems. If I am roughing out a part keeping the feed rates high enough to break the chip is possible. But when machining to finish sizes I can't do that and the chips pile up. I pack them into the paper barrels seen in the photo and

From the Chip Pan

Photo E2. Aluminum chips pile up.

take them to market. As I write this, in August of 2008, metal prices are the highest I have ever seen. I have paid less in the past for new material than I can get for scrap today. Where will it stop?